ArcGIS By Example

Develop three engaging ArcGIS applications to address
your real-world mapping scenarios

Hussein Nasser

PUBLISHING

BIRMINGHAM - MUMBAI

ArcGIS By Example

First published: August 2015

Production reference: 1240815

Published by Packt Publishing Ltd.
Livery Place
35 Livery Street
Birmingham B3 2PB, UK.

ISBN 978-1-78217-520-9

www.packtpub.com

Credits

Author
Hussein Nasser

Reviewers
Mohammed Alhessi

Nabil Allataifeh

Edward Hughes

Tram Vu Khanh Truong

Acquisition Editor
Nikhil Karkal

Content Development Editor
Aparna Mitra

Technical Editor
Mrunmayee Patil

Copy Editor
Sameen Siddiqui

Project Coordinator
Izzat Contractor

Proofreader
Safis Editing

Indexer
Mariammal Chettiyar

Production Coordinator
Conidon Miranda

Cover Work
Conidon Miranda

About the Author

Hussein Nasser is an Esri award-winning senior GIS solution architect working in the GIS field since 2006. He is the author of three books in the ArcGIS technology: *Administering ArcGIS for Server*, *Learning ArcGIS Geodatabases*, and *Building Web Applications with ArcGIS*, all by Packt Publishing. In 2007, he won the first place at the annual ArcGIS Server Code Challenge, conducted at the Esri Developer Summit in Palm Springs, California. In 2014, he started the IGeometry YouTube channel, where he periodically publishes educational GIS videos.

About the Reviewers

Mohammed Alhessi is a GIS professional and instructor who is interested in algorithms and applications of the geospatial theory. He has good experience in GIS analysis, development, and training. He has conducted numerous training courses for people from different backgrounds. The course topics include, but are not limited to, Enterprise Geodatabase Administration in MS SQL Server, spatial data analysis and modeling, and Python scripting for ArcGIS.

He worked at the University of Stuttgart as a GIS developer, programming geoprocessing tools using Java and Python. He is involved in many local GIS projects, providing consultancy for the local community. He is currently working as a lecturer at the Islamic University of Gaza, Palestine. He is also an instructor at University College of Applied Sciences in Gaza.

He has a bachelor's degree in civil engineering from the Islamic University of Gaza and a master's degree in geomatics engineering from Stuttgart University.

Nabil Allataifeh is a GIS specialist who holds a master's of engineering degree. He is a research assistant with more than 5 years of experience in GIS. He also has an experience in water resources engineering with a focus on hydrological modeling. He is currently a research assistant at the Department of Geography, working as a GIS specialist. He previously worked at the School of Engineering, University of Guelph as a research assistant/hydrological modeler. He has experience in ArcGIS 10.0 and 9.3, ArcMap, ArcScene, and ArcGlope. He also has experience in ArcGIS ModelBuilder, scripting, spatial analysis, and 3D analysis.

Edward Hughes has been working in the GIS industry since 1997. While performing professional roles at Western Power, he completed his degree in GIS at Curtin University.

He fist started data entry operations for the Western Power's Automated Mapping Facility Management (AMFM) system. Wind the clock forward to 2012 and Edward performed the role of an SME, upgrading the AMFM system to a corporate-based GIS system (Esri). Along the GIS pathway, he took up many opportunities within the GIS electrical utility sphere, as the project manager of a task force, where his work ranged from reporting bay model anomalies to automating a Vegetation Management System. He also recently played the key GIS component in implementing the Network Risk Management Tool (NRMT), which is based on Bayesian modeling.

He is a member of Mapping and Planning Support WA (MAPSWA), which provides mapping expertise to aid incident management teams at major emergency incidents, such as bushfires, urban storms, and tropical cyclones. A little coffee goes along way in these intense situations!

Tram Vu Khanh Truong is a transportation planner at the City of Greensboro and Greensboro Urban Area Metropolitan Planning Organization, North Carolina. She received her master's degree in regional and city planning from the University of Oklahoma. With more than 5 years of working experience in planning, she has experienced many facets of GIS, from using GPS to collect field data to analyzing GIS data and programming to automate processes. She possesses a strong passion for applying GIS in land use and transportation planning to support decision making.

Besides her enthusiasm for maps and numbers, she loves cooking, baking, reading, and having fun with her husband and adorable son.

Tram previously reviewed *Administering ArcGIS for Server* and *ArcGIS for Desktop Cookbook*, both by Packt Publishing.

www.PacktPub.com

Support files, eBooks, discount offers, and more

For support files and downloads related to your book, please visit www.PacktPub.com.

Did you know that Packt offers eBook versions of every book published, with PDF and ePub files available? You can upgrade to the eBook version at www.PacktPub.com and as a print book customer, you are entitled to a discount on the eBook copy. Get in touch with us at service@packtpub.com for more details.

At www.PacktPub.com, you can also read a collection of free technical articles, sign up for a range of free newsletters and receive exclusive discounts and offers on Packt books and eBooks.

https://www2.packtpub.com/books/subscription/packtlib

Do you need instant solutions to your IT questions? PacktLib is Packt's online digital book library. Here, you can search, access, and read Packt's entire library of books.

Why subscribe?

- Fully searchable across every book published by Packt
- Copy and paste, print, and bookmark content
- On demand and accessible via a web browser

Free access for Packt account holders

If you have an account with Packt at www.PacktPub.com, you can use this to access PacktLib today and view 9 entirely free books. Simply use your login credentials for immediate access.

For Nada

Table of Contents

Preface

Over the last two years, I have written three books on ArcGIS technology. Each book covers different topics and fields of this increasingly ubiquitous technology. Although I used examples and various real-life project approaches to explain the technology in all my books, this is the first book where the content evolves with the help of examples. I have been working with Esri ArcGIS since 2005 when ArcGIS 9.1 was released, so writing this title from a technological point of view was not difficult. In fact, it was thrilling. The challenging part was to come up with three unique real-life examples and to build them up as I wrote the book. Each example should target certain features of the technology and explain them along the way.

These three examples are all from my own imagination and they are not linked to or correlate with any actual projects that I personally worked on or witnessed. You will not find any of these examples in Esri's help or on any online resource. All the code that is available in this book is written from scratch for this book that you are holding in your hands.

The title of this book was designed for those who want to start using the ArcGIS technology or have been using it and want to learn more about how they can customize ArcGIS to do more. There are going to be three themes running throughout the book. The first theme covers *Chapter 2, App 1 – the Cell Tower Analysis Tool, Chapter 3, Mapping Signal Strength,* and *Chapter 4, Real-time Maneuvering,* which are tailored for beginners and developers. It features a Cell Tower Analysis Tool that displays a cell phone tower's signal range and signal strength on the map and shows you how cell phones connect—in simulated real time—to the tower with the strongest signal, all on top of ArcGIS for Desktop. You will learn ArcGIS add-ins for development.

The second theme covers *Chapter 5, App 2 - Extending ArcObjects, Chapter 6, Reviews and Ratings,* and *Chapter 7, Advanced Searching,* and is targeted at those who want to achieve more with ArcGIS. This theme features a restaurant mapping application that will be used to filter, search, and interact with restaurants on the map; it will also be used to view the reviews and the ratings of different users. You will learn how to write some ArcObjects code to work with geodatabases, query feature classes, and relationships. The last theme covers *Chapter 8, App 3 – Advanced ArcObjects, Chapter 9, Excavation Cost Calculation,* and *Chapter 10, Saving and Retrieving Excavation Designs* and is designed for those who are willing to try advanced programming. This theme features an excavation planning manager application. This application will propel the reader to the advanced stage, where they will write a real-life business-related deployable application. The Excavation Planning Manager helps construction workers plan their excavation for utilities and telecom networks beforehand in a given area and at an estimated cost of excavation. The application analyses the underlying soil type and green area to find out the cost of removing these areas by doing extensive spatial analysis. You will be able to store multiple designs of excavation and determine which is the cheapest or most applicable design. *Chapter 1, Getting Started with ArcGIS* ties all the chapters together and explains briefly what you will learn in all of them. It will also help you get started with the installations and will also tell you about the prerequisites.

In each of the themes, you will learn new features of ArcGIS and will be able to harness these features in your own code to enhance and extend ArcGIS capability.

What this book covers

Chapter 1, Getting Started with ArcGIS, introduces you to the book. Since you are new to ArcGIS, it will briefly explain what ArcGIS is and why a developer would customize ArcGIS to create cool applications with it. In this chapter, we illustrate each example, the technology, and the skills that a developer will acquire upon completing the example.

Chapter 2, App 1 – the Cell Tower Analysis Tool, kicks off with the first example, where you will learn how to develop on ArcGIS for Desktop using ArcGIS add-ins. Developers will write a tool to show a cell phone tower's signal range, display the strength signal on the map, and display how cell phones will connect—in simulated real time—to the tower with the strongest signal, all on ArcGIS for Desktop.

Chapter 3, Mapping Signal Strength, takes the application further to the next stage where you will learn about proximity tools, how to use them to measure distances between points, and perform analysis based on a result. This will help us in determining the closest tower, which will eventually be the one with the strongest signal. The signal strength can be calculated with the formula tower range-distance.

Chapter 4, Real-time Maneuvering, takes the application to a real-life scenario. In this chapter, we simulate a cell phone that moves on the map and switches towers for the best signal possible. The cell phone reads coordinates from a GPS textfile, which has been produced previously. The active tower will keep flashing while the cell phone is connected to that particular tower.

Chapter 5, App 2 – Extending ArcObjects, introduces our second application, the restaurant mapping application. You will create an application that will allow you to filter, search, and interact with restaurants on the map. This will also help you to view the reviews and ratings of different users. You will learn how to write some ArcObjects code to work with geodatabase, query feature classes, and relationships.

Chapter 6, Reviews and Ratings, introduces you to the relationship queries, which is a bit of an advanced topic that requires special care. You will be able to query related tables, such as reviews and ratings, pull this information, and display it on the application. A developer will learn how to highlight restaurants on the map by selecting it from the application.

Chapter 7, Advanced Searching, takes the application to a higher level with the advanced geodatabase search. In this chapter, we will introduce advanced spatial queries, where the user of the application will select an area and the application should display all the restaurants in the selected area according to their categories. You will also perform an advanced interface technique, where the developer will add a custom text box to the toolbar to search for restaurants and filter them accordingly as the user types in the box.

Chapter 8, App 3 – Advanced ArcObjects, will propel you to the advanced stage, where you will write a real-life business-related deployable application. The Excavation Planning Manager helps construction workers plan their excavation for utilities and telecom networks beforehand in a given area and at an estimated cost of excavation. The application analyses the underlying soil type and green area to find out the cost of removing these areas by carrying out extensive spatial analysis. You will be able to store multiple designs of excavation and determine which is the cheapest or most applicable design.

Chapter 9, Excavation Cost Calculation, will help you use advanced spatial operations to determine the estimated cost of a given excavation. The application will carry out spatial analysis on the area under the excavation polygon, and based on the soil type, the cost of removal of per 1 meter cube of soil might affect the overall cost of excavation. For instance, a stony area is more difficult to excavate than a regular sand area.

Chapter 10, Saving and Retrieving Excavation Designs, propels our application to the real-life scenario. Before this chapter, excavations were scattered and ungrouped; in this chapter, we will group excavations into designs. So here, a user can create a new design and add multiple excavations for his/her design and calculate the total cost of his/her design. A user will be able to search for a design, edit it, and delete it, along with all its underlying features.

What you need for this book

The following is a list of the tools that you'll need for this book:

- ESRI ArcGIS for Desktop 10.3, 10.2.*x*, or 10.*x*. This book uses ArcGIS 10.3. You can download a trial from `http://www.esri.com/ software/arcgis/ trial` or order it from your local ESRI distributor.

- Esri ArcObjects .NET SDK 10.3, 10.2.*x*, or 10.*x* that match the version of ArcGIS. The book uses ArcObjects SDK 10.3.

- Microsoft Visual Studio Express 2013, 2012, or 2010. This book uses the 2013 version, which you can download from `http://bit.ly/b04748_vs2013exp`.

- Microsoft .Net Framework 3.5 SP1, which can be downloaded for free from `http://qr.net/dnfm35sp1`.

Refer to *Chapter 1, Getting Started with ArcGIS*, for more details about the specific versions.

Who this book is for

Whether you are a student, GIS user, an analyst, or a programmer with basic or no knowledge of ESRI ArcGIS, this book is for you. This book assumes that you have basic programming skills in .NET technology.

Conventions

In this book, you will find a number of text styles that distinguish between different kinds of information. Here are some examples of these styles and an explanation of their meaning.

Code words in text, database table names, folder names, filenames, file extensions, pathnames, dummy URLs, user input, and Twitter handles are shown as follows: "Browse to the `Restaurants.gdb` geodatabase, select the `Food_and_Drinks` feature class."

A block of code is set as follows:

```
Dim pClosestTower As IFeature = pTowerLayer.FeatureClass.
GetFeature(closestTowerOID)
```

When we wish to draw your attention to a particular part of a code block, the relevant lines or items are set in bold:

```
Dim pTheTextElement As IElement = pTextElement
pTheTextElement.Geometry = pTextPoint
pDocument.ActiveView.GraphicsContainer.AddElement(pTheTextElement, 0)
pDocument.ActiveView.Refresh()
```

New terms and **important words** are shown in bold. Words that you see on the screen, for example, in menus or dialog boxes, appear in the text like this: "From the Visual Studio application, click on the **File** menu and then select **New Project**."

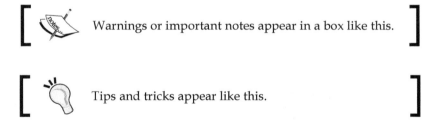

[Warnings or important notes appear in a box like this.]

[Tips and tricks appear like this.]

Reader feedback

Feedback from our readers is always welcome. Let us know what you think about this book—what you liked or disliked. Reader feedback is important for us as it helps us develop titles that you will really get the most out of.

To send us general feedback, simply e-mail feedback@packtpub.com, and mention the book's title in the subject of your message.

If there is a topic that you have expertise in and you are interested in either writing or contributing to a book, see our author guide at www.packtpub.com/authors.

Customer support

Now that you are the proud owner of a Packt book, we have a number of things to help you to get the most from your purchase.

Downloading the example code

You can download the example code files from your account at http://www.packtpub.com for all the Packt Publishing books you have purchased. If you purchased this book elsewhere, you can visit http://www.packtpub.com/support and register to have the files e-mailed directly to you.

Errata

Although we have taken every care to ensure the accuracy of our content, mistakes do happen. If you find a mistake in one of our books—maybe a mistake in the text or the code—we would be grateful if you could report this to us. By doing so, you can save other readers from frustration and help us improve subsequent versions of this book. If you find any errata, please report them by visiting http://www.packtpub.com/submit-errata, selecting your book, clicking on the **Errata Submission Form** link, and entering the details of your errata. Once your errata are verified, your submission will be accepted and the errata will be uploaded to our website or added to any list of existing errata under the Errata section of that title.

To view the previously submitted errata, go to https://www.packtpub.com/books/content/support and enter the name of the book in the search field. The required information will appear under the **Errata** section.

Piracy

Piracy of copyrighted material on the Internet is an ongoing problem across all media. At Packt, we take the protection of our copyright and licenses very seriously. If you come across any illegal copies of our works in any form on the Internet, please provide us with the location address or website name immediately so that we can pursue a remedy.

Please contact us at copyright@packtpub.com with a link to the suspected pirated material.

We appreciate your help in protecting our authors and our ability to bring you valuable content.

Questions

If you have a problem with any aspect of this book, you can contact us at questions@packtpub.com, and we will do our best to address the problem.

Getting Started with ArcGIS

Planning to build a product requires addressing a purpose and a goal. The product needs to either fix a problem or tackle a limitation that current solutions are unable to overcome. It might enter an existing market to compete with other products or it might define its own market if such a market doesn't exist. Once a set of problems to be solved are identified, a technology can then be used to build the product. Any selected technology comes with its' perks and limitations. The author of the product should be aware of them because they will eventually steer and shape his/her solution. This is what Esri tried to achieve with ArcGIS and what we will discover by the end of this book. In this chapter, we discuss the following topics:

- The history of ArcGIS
- An introduction to ArcGIS for Desktop
- Customizing ArcGIS for Desktop
- App 1 – the cell tower analysis tool
- App 2 – the restaurant mapping application
- App 3 – the excavation planning manager

The history of ArcGIS

Esri, Environmental Systems Research Institute, knew there was a starving market for location-based systems also known **geographic information systems (GIS)**. In 1990s, Esri started working on a product that later became one of the best enterprise solutions for GIS implementations on Windows systems. In 1999, ArcGIS was released. Since then, ArcGIS has become the most used commercial GIS solution. ArcGIS was then renamed **ArcGIS for Desktop**, and the ArcGIS name was used as a product line instead to carry lots of products under it.

When the Web started to become ubiquitous in early 2000s, Esri adopted the Web by rolling in **ArcGIS for Server** and gradually ArcGIS functionalities as web services so that it could be supported on multiple platforms including mobile phones.

A decade later when the cloud solutions began to surface, Esri released its **Software as a Service (SaaS)** solution **ArcGIS Online**. Designed to simplify the user experience, ArcGIS Online hides all the ArcGIS "contraptions" and technologies to relieve the user from maintaining the hardware and software, leaving the user to do what they do best, mapping. Having everything in the cloud allows users to focus on their work instead of worrying about configurations, spinning up servers and databases, and running optimization checks.

 SaaS, a cloud-based software distribution model where all infrastructure, hardware, management software, and applications are hosted in the cloud. Users consume the applications as services without the need to have high-end terminal machines.

Today, Esri is pushing to enhance and enrich the user experience and support multiple platforms by using the ArcGIS Online technology.

In this book, we target one of the core products of the ArcGIS family — ArcGIS for Desktop. By using real-life examples, we will demonstrate the power and flexibility of this 16+ year-old product ArcGIS for Desktop. We are going to use the various tools at our disposable to show how we can extend the functionality of ArcGIS for Desktop.

An introduction to ArcGIS for Desktop

In this section, we will talk about ArcGIS for Desktop: What is it? How does it work? What different components does it consist of? What does it require to run? We will also explain about core ArcGIS concepts and will use the application out-of-the-box.

ArcGIS for Desktop was originally designed to allow users to author maps and spatial data. The ability for analysis was added to this product to make it one of the best GIS desktop solutions on the market. ArcGIS for Desktop consists of many components. Firstly, ArcMap is the map authoring and viewing tool, and this is the one we will be dealing with throughout this book. You can run tools on your map, edit, analyze, or export your map to different formats to support other platforms. The second component is ArcCatalog. You can use it to connect to geodatabases, author your own geodatabases, manipulate datasets, feature classes, and much more. We will be defining the ArcGIS geodatabase in the coming sections. You can learn more about geodatabases in my other book *Learning ArcGIS Geodatabases, Packt Publishing*. There are other products that come under the umbrella of Desktop like ArcGlobe and ArcScene for 3D analysis, which are out of the scope of this book.

ArcGIS for Desktop licenses

ArcGIS for Desktop has three different licenses: **Basic**, **Standard**, and **Advanced** previously known as **ArcView**, **ArcEditor**, and **ArcInfo**, respectively. The Basic license mainly gives you the viewer features, which allows you to read map documents and query the data. You can, in fact, do simple editing with the Basic license, but it is very limited. The Standard license is the editor, which allows you to view, create, and edit maps and spatial data. It allows you to edit and create complex data structures and allows multiple users to edit the same geodatabase. The Advanced license allows you to do what the Basic and Standard do, plus the ability to do advance data analysis and modeling, which we will not require in this book. You can take a look at the differences in details at `http://bit.ly/b04847_agslicenses`.

In this book, the first two examples only require the Basic license. However, the third example requires the Standard license to fully implement it. Esri provides the Standard license for 60 days, which you can get by creating an account at `http://www.esri.com/`.

The system requirements of ArcGIS for Desktop

The latest out-of-the-box ArcGIS for Desktop can be downloaded from the official Esri website at `http://bit.ly/b04748_agsfree`. However, if you want to customize, in the same way we will be doing in this book, you should officially request the media disc from your local Esri distributor that will have the **ArcObjects** SDK.

 ArcObjects is a software development kit by ArcGIS that can be used by software developers to extend the ArcGIS functionality.

ArcGIS for Desktop requires the .NET Framework 3.5 service pack 1 and Microsoft Internet Explorer 9.0 or higher in order to run. The .NET Framework can be downloaded from `http://bit.ly/b04748_dotnet35`. Some operating systems, such as Windows Server can be configured to enable the .NET Framework, instructions to do that can be found in the same link. The system requirements for running ArcGIS for Desktop as of version 10.3 and full details on the system and hardware requirements can be found at `http://bit.ly/b04748_ags103sysreq`.

In this book, I will be using Microsoft Windows 8.1 Pro with ArcGIS for Desktop 10.3. Feel free to use any version of Desktop (10 or higher) with the supported version of Windows as per the system requirements in the following table:

Product Version	Supported OS	Reference
ArcGIS 10.3	Windows 7, 8, 8.1, Server 2008, Server 2008R2, Server 2012, Server 2012R2	`http://bit.ly/ b04748_ags103sysreq`
ArcGIS 10.2.*x*	Windows XP, Vista, 7, 8, 8.1, Server 2003, Server 2008, Server 2008R2, Server 2012, Server 2012R2	`http://bit.ly/ b04748_ags102sysreq`
ArcGIS 10.1	Windows XP, Vista, 7, 8, Server 2003, Server 2008, Server 2008R2, Server 2012	`http://bit.ly/ b04748_ags101sysreq`
ArcGIS 10.0	Windows XP, Vista, 7, Server 2003, Server 2008, Server 2008R2	`http://bit.ly/ b04748_ags10sysreq`
ArcGIS 9.3.*x*	Windows XP, Vista, 7, Server 2003, Server 2008, Server 2008R2	`http://bit.ly/ b04748_ags93sysreq`
ArcGIS 9.2.*x*	Windows XP, Vista, Server 2000, Server 2003	`http://bit.ly/ b04748_ags92sysreq`

The examples in this book can also be applied to older versions of ArcGIS (10.0, 10.1, 10.2.*x*). I will be providing designated copies of the data and map documents for each version so that you can freely work with the version of ArcGIS you prefer.

 ArcGIS versions prior to 10 won't be able to take advantage of the new add-in feature.

The important concepts of ArcGIS for Desktop

Before we dive into customizing ArcGIS, it is important to know some key concepts and definitions. We will start with the geodatabase.

The ArcGIS geodatabase

The database is a fascinating storage system. It allows you to retrieve, store, and edit the different types of information such as text, images, music, and videos. However, for people who work with maps, we feel there is a missing element in that compound, that is, location. Adding location information to database helps applications bring life to the tabular records in the database and make it available visually. Esri has done this in its ArcGIS product and called this special location-based database a geodatabase.

The ArcGIS geodatabase is the proprietary database for Esri. All Esri geospatial software is built around this geodatabase.

Adding location information to a database requires two parameters: the actual location coordinates and how these coordinates are supposed to be drawn, which is also known as the **spatial reference**. The spatial reference describes whether the location is projected on to a two- or three-dimensional map, and it should be defined for every dataset in the geodatabase that has a spatial component. While working in ArcMap, all datasets should share the same spatial reference.

 A spatial reference is a collection of properties that describes the system for locating a particular object in a coordinate system. You can find more information about this topic at http://bit.ly/b04748_spatialref.

There are a lot of spatial references tailored for different locations on the earth. There are some standard references used universally, and among them is the WGS 84, which we will be continuously using in this book.

Let us start using the software and get familiar with geodatabase components. Make sure you have installed ArcGIS for Desktop and then follow these steps:

1. First of all, we want a geodatabase to work with. Create a new folder in your root drive c:\ArcGISByExample\. In the supporting files for this chapter, copy the B04847_01_Files folder to the C:\ArcGISbyExample folder.

2. From the Start menu, locate and run ArcCatalog 10.3 (or your version of ArcCatalog). It is the one with the cabinet icon.

 You can dock and pin ArcCatalog in your start menu to access it quickly.

3. From the **Catalog Tree** window, right-click on **Folder Connections** and click on **Connect To Folder**. This will establish a connection with the folder that contains the geodatabase.

4. From the **Connect To Folder** dialog, browse and select the
   C:\ArcGISbyExample folder and click on **OK**, as illustrated in the
   next screenshot. Note that if you don't see the **Catalog Tree** window,
   you can show it from the Windows menu in the toolbar.

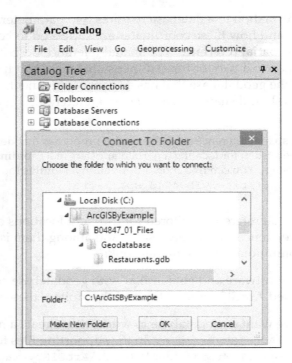

5. You should see that the C:\ArcGISbyExample folder has been
   added to the Folder Connections folder. Use this folder to browse to
   C:\ArcGISByExample\B04847_01_Files\Geodatabase\Restaurants.gdb,
   as shown in the following screenshot:

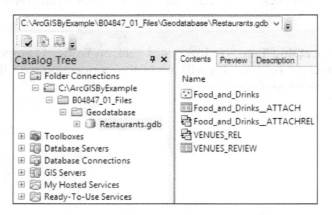

6. Make sure that the **Content** tab is active. You should see the different objects that this geodatabase consists of. The first object is `Food_and_Drinks`, which is the feature class of some restaurants. The `Food_and_Drinks` object has a one-to-many relationship with `VENUES_REVIEW` which stores the reviews of a given restaurant.

[The feature class is one of the basic objects in a geodatabase. This object is a table with a shape attribute that defines the location and geometry. It could be a point, line, or a polygon.]

7. You can view the content of the feature class by selecting it and clicking on **Preview**, as shown in the following screenshot. The default preview is **Geography**, which visually displays the points:

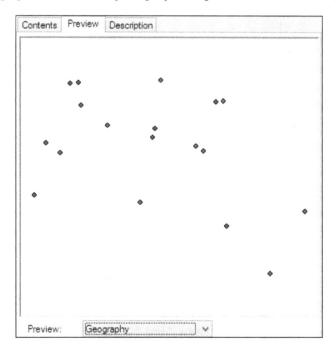

8. You can also display a tabular view by changing the **Preview** type to **Table**, as illustrated in the following screenshot:

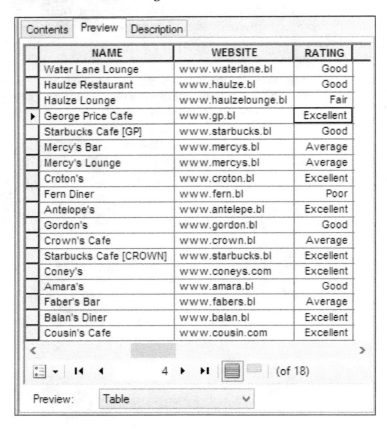

9. Go back to the **Contents** tab, from **Catalog Tree**, right-click on the `Food_and_Drinks` feature class and select **Properties**.

10. Activate the **General** tab and take a look at Alias Name, we can freely change this name without actually changing the physical feature class name for convenience reasons. The current Alias Name **Food and Drinks Venues in Belize** is quite long, so let us change it to `Venues` only. Click on **Apply** to save your changes.

11. Activate the **Fields** tab, which is the columns or attributes that this feature class consists of. Take note of the different data types for each field. Pay attention to the **SHAPE** field, which is created by default and the data type is **Geometry**.

12. Activate the **Subtypes** tab; here we can define multiple types for our feature class. In our case, we have five different restaurant types number coded.

13. Click on **OK** to close the feature class properties.

14. Close ArcCatalog.

Working with the map layers

Now that we have worked with ArcCatalog and learned about the basics of the geodatabase, it is time to learn about the map:

1. From the Start menu, locate and run ArcMap 10.3 (or your version of ArcMap). It is the one with the map and lens icon.

2. If you are opening ArcMap for the first time, you will be prompted with the getting started dialog. Click on **Cancel** to work on the default document.

3. We want to work with our geodatabase on ArcMap. To do that, we need to add a feature class to the map.

4. From the **Table of Content** window, right-click on the **Layers** node and click on **Add Data**. This will open a dialog to select a geodatabase.

5. Since we established a folder in ArcCatalog, you should see it in **Folder Connection** under the **Look In** dropdown.

6. Browse to the Restaurants.gdb geodatabase, select the Food_and_Drinks feature class, and then click on **Add**, as illustrated in the following screenshot:

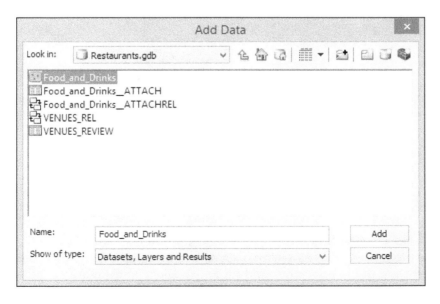

7. You will see that a new layer has been created under layers named **Venues**.
   This is the representation of the feature class. You can see that the name of
   the layer is actually the alias name of the feature class by default, which we
   have renamed in the *The ArcGIS geodatabase* section. ArcMap creates this
   layer wrapper to visual a feature class, change symbology, control labels,
   scaling, and so many other things.

A layer is an ArcMap object and a visual representation of a physical
feature class. A layer does not exist by itself and needs a source dataset
to read data from.

A symbology is a notation for the features in a feature class. A given
feature class might have multiple symbologies based on its attributes.

8. Note that different symbologies have been assigned based on the restaurant
   subtypes that we have mentioned in the *The ArcGIS geodatabase* section. See
   the following screenshot:

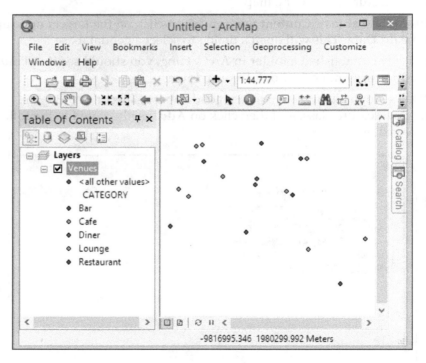

9. We can change the symbology to make it more relevant; click on the point
   next to **Bar** to change its symbol. This will bring up the **Symbol Selector**.
   Type Bar in the search box and hit *Enter*. Select your favorite symbol and
   click on **OK**, as shown in the next screenshot:

10. You should see that the map has been refreshed with the new symbology, as shown in the next screenshot. You can see how rich you can make your map by using these built-in tools. Imagine what we can do if we could extend this to the next level, as we will see in the next chapter.

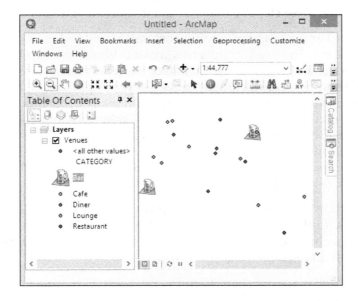

11. Close ArcMap and choose not to save your changes.

Customizing ArcGIS for Desktop

In this section, we will discuss the benefits of customizing ArcGIS. When a particular requirement or feature is not available in ArcGIS, we can actually extend the ArcGIS functionality to do that for us. In this book, we will explain two different approaches for deploying ArcGIS customization respectively: add-ins and extending ArcObjects. You will need the ArcObjects SDK to start the development. There are many other ways for customizing ArcGIS, including user interface customization using **Visual Basic for Applications** (**VBA**), modeling and scripting using Python, and building standalone applications using ArcGIS Engine. In this book, we will use the add-ins and extending ArcObjects method.

 ArcGIS add-ins is a building approach where the developer customizations are categorized and controlled by ArcGIS. Add-ins can be disabled and enabled by the user of ArcGIS at any given time.

ArcGIS for Desktop comes with great set of built-in tools that can help you solve interesting mapping problems. However, there comes a time where your problem is a complex one. This is where you might need to extend and customize the functionality of ArcGIS to provide a suitable solution to your problem. The examples in this book require customizations to tackle them. In this section, we discuss the different customization approaches to set up our development environment. Another reason to extend ArcGIS, for example, a certain functionality, might be available in ArcGIS, but you need to perform 10 or 15 steps to achieve it, and customizing the product can group and automate these steps so that you can default all of them in a few clicks.

The first attempted approach to providing customization for ArcGIS was through VBA. This is similar to the macroscripts in Microsoft Word and Excel. You could write an application and save it in the map document and later share this document, and the person running your document could use your application. It was a convenient approach for sharing mapping, but with many problems. The main problem was the security. The document might contain malicious code that would execute with user privileges and can potentially harm the user. That is why this approach was discouraged and has been replaced with ArcGIS add-ins and the extensions building approach. Today, you can still develop using VBA by installing the VBA compatibility setup.

Customizing ArcGIS for Desktop requires that you either build add-ins or use the classical **Dynamic Link Library** (**DLL**) approach and register it with ArcGIS for it to work. Both approaches use ArcObjects as the underlying technology, however, the final building technique is different.

These built approaches are not share-friendly, however, Esri came up with a beautiful solution and platform for sharing, and that is ArcGIS Online. That discussion though should be an entirely different book.

The system requirements of ArcObjects

In order to customize ArcGIS for Desktop, we will require installing some more components. Microsoft Visual Studio, which will be our **Integrated Development Environment (IDE)**. This is where we will be writing code in .NET to customize ArcGIS.

The second component that we will also need to install is ArcObjects SDK for .NET, which will add the software development kit and Visual Studio plugins to write Desktop applications.

> IDE is a software that allows computer programmers to write code in order to develop software. The IDE usually consists of a source code editor, syntax highlighter, compiler, builder, and a debugger, which help the programmer in the software development process.
>
> It is important to install ArcObjects SDK after completely installing Visual Studio so that ArcObjects will be able to install plugins on top of the Visual Studio IDE.

In this book, we will be using Microsoft Visual Studio 2013 Express for Windows Desktop as our IDE. The software can be downloaded for free from the official Microsoft website at `http://bit.ly/b04748_vs2013exp`. This is quite a big download and it will take some time depending on your Internet connection. When you install ArcObjects SDK, the setup will detect your current Visual Studio and install the plugins accordingly, as shown in the following screenshot:

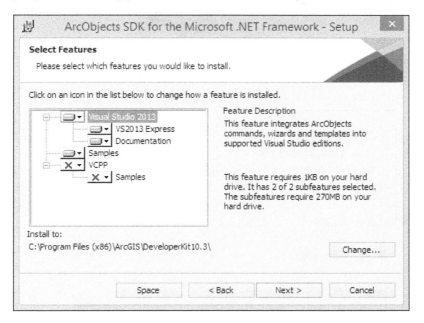

You can also use Visual Studio 2012 with ArcObjects SDK 10.3. Just make sure to install the Visual Studio before you install the ArcObjects SDK for .NET. For a complete list of system requirements for ArcObjects SDK 10.3, follow `http://bit.ly/b04748_ao103sysreq`.

If you are programming under different system configurations, take a look at the following table to make sure you comply with the system requirement:

Product version	Supported IDE	Reference
ArcObjects SDK 10.3	VS2013, VS2012	`http://bit.ly/b04748_ao103sysreq`
ArcObjects SDK 10.2.x	VS2012, VS2010	`http://bit.ly/b04748_ao102sysreq`
ArcObjects SDK 10.1	VS2010	`http://bit.ly/b04748_ao101sysreq`
ArcObjects SDK 10.0	VS2008, VS2010	`http://bit.ly/b04748_ao10sysreq`
ArcObjects SDK 9.3.x	VS2008, VS2010	N/A
ArcObjects SDK 9.2.x	VS2005, VS2008	N/A

I couldn't find official online references to support my claims for 9.3.*x* and 9.2.*x*, but from a personal experience, I did use ArcObjects with 2005 and 2008 on both the 9.2 and 9.3 systems and it was working flawlessly.

Verifying the installation of ArcObjects

In this section, we will validate the ArcObjects SDK installations and Visual Studio. To make sure that you have installed ArcObjects SDK correctly, follow these steps:

1. From the start menu, locate and run VS Express 2013 for Desktop. If you are using a different version of Visual Studio, refer to the system requirements section to make sure your version complies with ArcGIS.

2. From the Visual Studio application, click on the **File** menu and then select **New Project**.

3. From the **New Project** dialog, expand the **Templates** node, and then expand **Visual Basic**. This is the language we will be using in this book; you can freely use C# if you would like. If you have installed ArcObjects SDK successfully, you should see the ArcGIS node where we have all the different approaches for customizing ArcGIS. This is illustrated in the following screenshot:

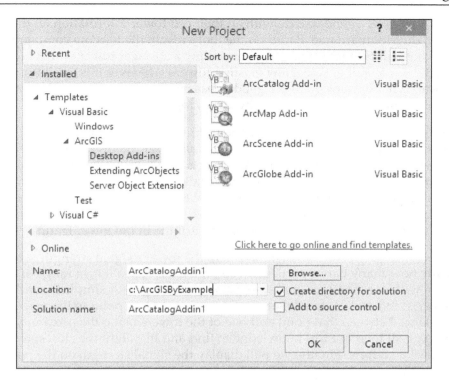

4. Close Visual Studio.

App 1 – the cell tower analysis tool

In this section, we introduce the first example that we will work on. A telecom company wants to measure their user experience when it comes to cell phone signal reception. The tool you will write will help the company decide whether to add more towers, relocate, or upgrade existing towers to provide maximum signal coverage so that users can experience uninterrupted reception while using their cell phone.

TelZaViBa is a telecom company that provides cell service for their customers. Recently, some customers have been experiencing a weak signal on the Boulevard du Montparnasse. To analyze the situation, TelZaViBa had to analyze their cell towers in that area. So they asked us to write a tool on ArcGIS for Desktop that simulates a person with a cell phone walking in the Boulevard du Montparnasse. The tool should show the current signal strength at all time and record the weakest signal spots by highlighting it on the map.

TelZaViBA gave us a geodatabase with all their cell tower information on the Boulevard du Montparnasse. Based on the tool's result, the telecom company can then do what will be necessary, such as installing a stronger tower with a higher range in the weak spot or relocating existing towers wherever it is feasible and economical. What we have here is a geodatabase with information, and we need to take this information to the next level by analyzing it.

This application will span into three chapters. Since this is the first example, we will spend some time in *Chapter 2, App 1 – the Cell Tower Analysis Tool*, to get you familiar with ArcGIS add-ins and the ArcObjects interfaces before we dive into the development. We will prepare the ground by talking about layers, feature classes, features, and geometry. We will then learn how to do some topological operations on the geometry to draw the signal range buffer. Then we will draw the signal range based on the radius value that is stored as an attribute in the tower feature class.

In *Chapter 3, Mapping Signal Strength*, we map the tower's signal strength, which is basically how many bars a particular cell phone has when it is in range of a cell tower. We will measure the signal strength in percentages for simplicity and then we will convert it to bars. To do all that, we first need to add a point to the map and then find a distance between that point and one of the towers using the proximity tools in ArcGIS. We will then use this knowledge to find and highlight the closest tower to the point we just drew. Finally, we will display the signal strength on the point using ArcGIS graphics.

In *Chapter 4, Real-time Maneuvering*, we do the real-time maneuvering and things get interesting. We will simulate a person walking along the boulevard with a cell phone and then use the logic we wrote in *Chapter 3, Mapping Signal Strength*, to establish the signal strength and the closest tower. We will simulate this by reading the previously recorded text file **Global Positioning System (GPS)** points and load them into our tool. With each step the signal will get updated with the new value based on our signal calculation algorithm. The active connected tower will be blinking on the map along with the cell phone.

 GPS provides the location and time information using satellites on the earth. Nearly all new smart phones are equipped with GPS receivers that can identify the device's location with respect to the earth.

TelZaViBa can use this tool to simulate cell phones and monitor the signal strength on the boulevard and find the weak signal spots.

This example is not an actual project and is not related to any country whatsoever. All data, maps, and ideas in this example are my sole creation and have not been copied or repurposed from other resources.

App 2 – the restaurant mapping application

This is the second example that features a restaurant mapping application where you will build an application on top of ArcGIS for Desktop that allows users to view, search, highlight restaurants on the map, and compare their ratings and reviews.

Belize is thriving with tourism. Lots of tourists go there on holidays to enjoy its beautiful beaches and a wide range of restaurants. The government of Belize is trying to enrich the tourists' experience by finding their favorite restaurants in the country more effectively.

To accomplish that, a new project titled *Bestaurants* has been proposed to design a restaurant mapping application on top of ArcGIS for Desktop to feature the best restaurants in Belize. The application will contain a map that shows the city of Belize and the restaurants with key icons based on the restaurant type. For example, a café will be shown as a coffee mug and a restaurant will be displayed as a fork and a knife. Users should be able to search for restaurants by name, region, category, or rating.

Using the Bestaurants geodatabase, we will build this application from scratch to satisfy these requirements. In *Chapter 5, App 2 – Extending ArcObjects*, we start off by learning a bit about a different deployment approach with ArcObjects. We learn about building our first toolbar on ArcMap and add a button to it. We will learn how to query the subtypes that we have mentioned in the ArcGIS geodatabase section and populate them in a form. We will then filter the restaurants based on a selected subtype. We will kick off *Chapter 6, Reviews and Ratings*, by talking about relationships and then use this knowledge to learn how to query relationship classes in ArcObjects. This will be useful to retrieve ratings and reviews since these are related information that is located in another table. We will also learn how to filter the layer to show searched results. Finally, in *Chapter 7, Advanced Searching*, we include the region feature class and use the intersect tool to find all restaurants within a region and populate them in the form. We will also show how to add a text box to our toolbar and add search functionality to search for results as we type in the text box. This will make it easy for users to simply type the name or even part of the name of the restaurant and show it on the map.

This example is not an actual project and is not related to any country whatsoever. All data, maps, and ideas in this example are my sole creation and have not been copied or repurposed from other resources.

App 3 – the excavation planning manager

The excavation planning manager is a tool that we will be writing to help construction designers plan their excavation for utilities and telecom networks. The application analyzes the underlying soil type and green area to find out the cost of removing these areas by doing extensive spatial analysis and editing. Note that we will require the Standard license for this example.

When utility and telecom companies want to lay out their underground assets, cables, and pipes, they need to excavate the ground first. This is a challenging task since there are different types of soil and each has special kind of machinery and equipment, and, therefore, cost.

YharanamCo is a construction contractor experienced in executing efficient and economical excavations for utility and telecom companies. When YharanamCo's board of directors heard of ArcGIS technology, they wanted to use their expertise with the power of ArcGIS to come up with a solution that could help them cut costs even more. Soil type is not the only factor in excavation; there are many factors including the green factor, where you need to preserve the trees and green area while excavating for visual appealing. Using ArcGIS, YharanamCo can determine the soil type and green factor and calculate the cost of an excavation.

The excavation planning manager is the application you will be writing on top of ArcGIS. This application will help YharanamCo create multiple designs and scenarios for a given excavation. This way they can compare the cost for each one and how many trees they could save by going through another excavation route. YharanamCo has provided us with the geodatabase of the soil and trees data for one of their new projects for our development.

In *Chapter 8, App 3 – Advanced ArcObjects*, since we will create excavations on the map, we will learn the geodatabase editing. We will then add a tool to draw polygons using ArcObjects drawing tools. Then we will view and edit the excavations that we created.

In *Chapter 9, Excavation Cost Calculation,* the actual advanced spatial analysis and cost estimation happens. We will write this cost calculation module that uses the soil and trees layers and excavation. This is why we require the Standard license to perform such advanced spatial analysis.

In *Chapter 10, Saving and Retrieving Excavation Designs,* we will propel our application. We will group multiple excavations into a design. We will then allow the user to create multiple designs. The user can open, close, edit, compare, and delete designs. Each design will be a dedicated geodatabase; therefore, we will be making copies and dealing with multiple geodatabases at once. It will be a great experience.

Summary

In this chapter, you learned about the different components of ArcGIS for Desktop, ArcMap, and ArcCatalog. You used ArcCatalog to learn more about ArcGIS like geodatabase, spatial references, and feature classes. You also used ArcMap to add a layer and change its symbology. After paving the way with these ArcGIS basic concepts, you were briefly introduced to the three examples that you will be working on through the course of this book. The first example talked about the basic spatial customization. The second one taught you intermediate skills for working with the geodatabase. The last example featured advanced geodatabase and mapping techniques that combined will set you up to take your ArcGIS development skills to the next level.

In the next chapter, you will learn how to develop using ArcGIS add-in for your first example, the TelZaViBa cell tower analysis tool.

2
App 1 – the Cell Tower Analysis Tool

TelZaViBa is a telecom company that provides cell service for their customers. Recently, some customers have been experiencing weak signals in the Boulevard du Montparnasse in Paris, France. To analyze the situation, TelZaViBa has to analyze their cell towers in that area. So they asked us to write a tool on ArcGIS for Desktop that simulates a person with a cell phone walking in the Boulevard du Montparnasse. The tool should show the current signal strength at all time and record the weakest signal spots by highlighting them on the map.

In this chapter, we will discuss the following topics:

- An introduction to ArcGIS add-ins
- Preparing the TelZaViBA data and project: showing tower's range
- Drawing signal range based on attribute values

An introduction to ArcGIS add-ins

In this section, we will introduce the concept of the ArcGIS add-ins building approach. We will then build our first Hello, ArcGIS project, which will help us get started with ArcGIS add-ins. The Hello, ArcGIS project will be a small example that explains the basics of add-ins. As explained in the previous chapter, there are different approaches to building applications on ArcGIS for Desktop. The first and oldest method is **Visual Basic for Applications** (**VBA**), the second one is add-in, and the third one is extending ArcObjects, which we will be using in the last two examples.

The reason I started with add-ins is because it is the easiest and most convenient method for development and deployment. When you write and deploy an add-in, ArcGIS detects it and asks you whether you want to install it or not. Add-ins can be later disabled or enabled based on need, which makes them a secure approach for development.

Creating the Hello, ArcGIS add-in project

In this section, we will show how to create an add-in project using Visual Studio. To create an add-in project for ArcGIS, follow these steps:

1. From the Start menu, locate and run Microsoft Visual Studio.

2. From the Visual Studio application, point to the **File** menu and then click on **New Project**.

3. Under the **Templates** node, expand **Visual Basic**, this is the language we will be programming with, then expand ArcGIS, and click on **Desktop Add-ins**. You can write add-ins for all ArcGIS for Desktop products as you can see. We want to write add-ins on top of ArcMap, so select **ArcMap Add-ins**.

4. In the **Name** field, type the name of the project `HelloArcGIS` and point the location to the `C:\ArcGISByExample\HelloArcGIS` folder, as illustrated in the following screenshot:

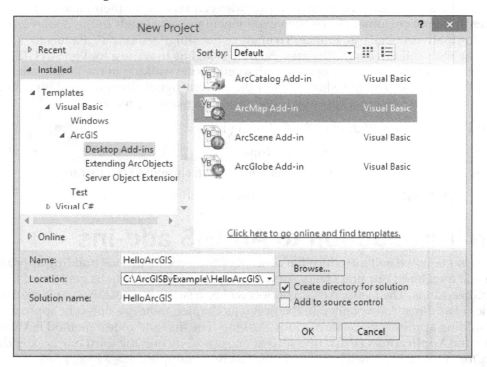

5. Click on **OK**, this will show the **ArcGIS Add-in Wizard**.

6. We can use the wizard to autogenerate our controls and other codes, but we are going to do that manually, so simply click on **Finish** to create the project.

7. Once the project has been created, you will see two files: `config.esriaddinx` and an image as an icon for your add-in. The `config.esriaddinx` file contains metadata about your project like the name and description. To add a functionality to our project, we need to add a control to do this, point to the **Project** menu and then click on **Add Class**.

8. From the **Common Items** node expand the **ArcGIS** node and then click on **Desktop Add-ins**. Click on **Add-in Component** and call it `btnHelloArcGIS`, and then click on **Add**.

9. From the **Add-in Wizard**, select **Button**, for the button caption type `Hello ArcGIS` and then click on **Finish**.

10. This adds two files: a class and an image. Double-click on the `btnHelloArcGIS` class.

11. Note that this class inherits from a base class `ESRI.ArcGIS.Desktop.AddIns.Button`, which will give it the functionality of a button. What is interesting for us here is the `onClick` event, which is the code that will get executed when one clicks on the button. Write the following code in the `onClick` event:

```
MsgBox("Hello, ArcGIS!")
```

For more details, take a look at the following screenshot:

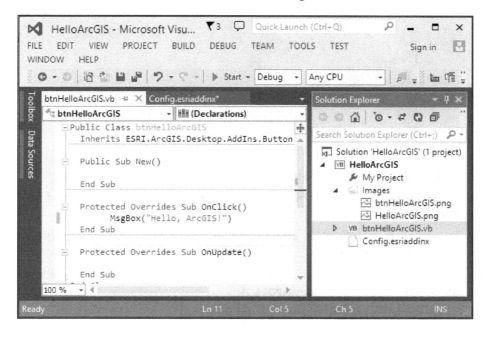

12. It is time to build our project and see the fruits of our work. Point to the **Build** menu and then click on **Build solution**. You should get a **Build succeeded** status message if you have written the code correctly.

13. Run ArcMap.

14. From the **Customize** menu, click on **Customize Mode**.

15. Activate the **Commands** tab and then type `Hello ArcGIS` in the search box, as illustrated in the following screenshot. Drag the **Hello ArcGIS** button near the **Zoom in** button to add it to the toolbar.

16. Close the **Customize** dialog.

17. By clicking on your new button, you will notice that the message Hello ArcGIS is displayed.

18. To add your own toolbar, add a new class to your project and choose **Add-in Command Container**.

19. Select **Toolbar** from the wizard and then select your button to be added to the toolbar from the drop-down list.

20. Close ArcMap and then close Visual Studio.

Preparing the TelZaViBa data and code

Here we will explain the concepts of the geodatabases such as layer, feature class, feature, and geometry. We will also prepare the TelZaViBa geodatabase, map document, and code.

Before we dive into customizing ArcGIS, it is important to know some key concepts and definitions. We will start with the geodatabase. We will create and prepare the TelZaViBa geodatabase from scratch and we will populate it with some towers. Then we will prepare the map document on which we will work. Finally, we will start writing some code.

 The ArcGIS geodatabase is the proprietary database for Esri. All Esri geospatial software is built around this geodatabase.

Note that you can skip the geodatabases and map preparation part and start working with the exercises. However, it is recommended that you go through the preparation part if you are new to ArcGIS as it will teach you how to create geodatabases and map documents. You can find the final fully prepared geodatabases and map document under the supporting files of this chapter. Also note that you will need at least a Standard license in order to work on the geodatabases and map preparation sections since they involve editing.

Preparing the geodatabase

In this section, we will use ArcCatalog to create the geodatabase. ArcCatalog is an application that allows us to author geodatabases, browse through them, add and delete datasets from a geodatabase, and so on. To create your first geodatabase, perform the following steps:

1. Locate and run ArcCatalog; you can find it in the Start menu under the `ArcGIS` folder, as explained in the previous chapter.

2. Once you start the application, make sure you can see the **Catalog Tree** window. This is the folder view of your computer and where we will be doing most of the work.

3. To show **Catalog Tree**, point to the **Windows** menu and then click on **Catalog Tree**.

4. Next, you need to specify the folder where you will be creating your TelZaViBa geodatabase.

5. Minimize ArcCatalog and create a new folder named `C:\ArcGISByExample\telzaviba` using Windows Explorer. This is where the geodatabase will go.

6. From **Catalog Tree**, right-click on the **Folder Connections** node and click on **Connect to Folder**. This feature allows you to connect to your `Windows` folder.

7. Browse to the new folder you just created, `C:\ArcGISByExample\telzaviba`, and then click on **OK**.

8. Select the folder and then right-click on the empty view to the right, point to **New**, and then click on **File Geodatabase**, as shown in the following screenshot:

9. This will create a file geodatabase in the specified folder with the default name of `Geodatabase.gdb` and rename the new geodatabase to `telzaviba`. The `.gdb` extension is automatically appended.

Now that we have created an empty geodatabase, it's time to create our first dataset. The feature class will store the towers information.

The feature class is one of the basic objects in a geodatabase. This object class is a table with a shape attribute, which could be a point, line, or a polygon.

To create the `Towers` feature class, follow these steps:

1. Click on the `telzaviba` file geodatabase, right-click on the empty right panel, point to **New**, and then click on **Feature Class**, as shown in the following screenshot:

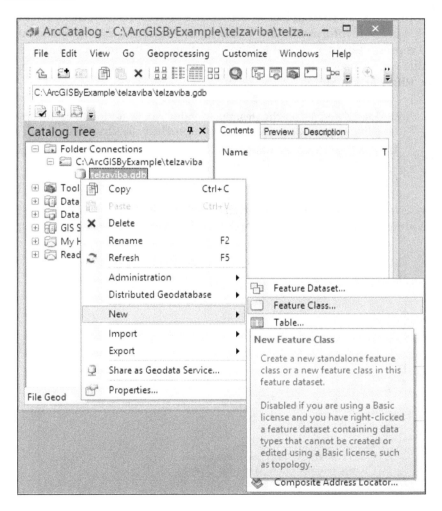

2. This will open up the **New Feature Class** dialog, where you can specify the basic properties of the feature class. In the **Name** field, type `Towers`. This is the physical name of the feature class in the geodatabase, and it should not contain special characters, or spaces.

3. Type `Cell Towers` in the **Alias** field. This is a description of the feature class name. It can be anything you want. When you add a new layer, it takes this alias name by default.

4. In the **Type** drop-down list, select **Point Features** to create the feature class with point geometry. The **Geometry Properties** section offers advanced options that can be enabled on the feature class. This includes the M value that helps in route information for linear features and the Z value that is used for 3D representation and enables the elevation and extrusion of features. The Z value can be useful, for example, if a restaurant is located on the 11th floor of the Ritz Carlton Hotel.

Besides X and Y coordinates, the M value can be added to each vertex on a line to provide more information such as the direction.

Unlike X and Y coordinates, the Z value can be considered as a height of a feature upward or downward. The value can be assigned to features so they are represented in a 3D space.

5. Since we won't need store route data or 3D data at this stage, leave the M and Z values unchecked. Click on **Next**, as illustrated in the following screenshot:

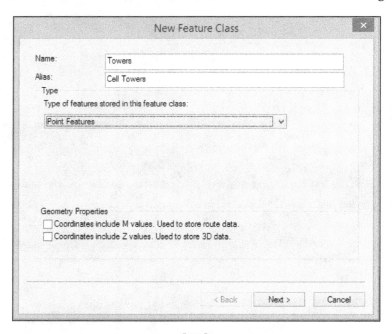

In the next dialog, you will set the spatial reference for our new feature class. You will use the WGS_1984_Web_Mercator standard spatial reference, which is also used by Google Maps. The spatial reference has been explained in *Chapter 1, Getting Started with ArcGIS*.

6. In the spatial reference drop-down list, type WGS_1984_Web_Mercator and press *Enter* to find the item.

7. Expand the **Projected Coordinate Systems** node and then the **World** node, and then click on the **WGS_1984_Web_Mercator (auxiliary sphere)**, as shown in the following screenshot.

8. Click on **Next** to move to the next form:

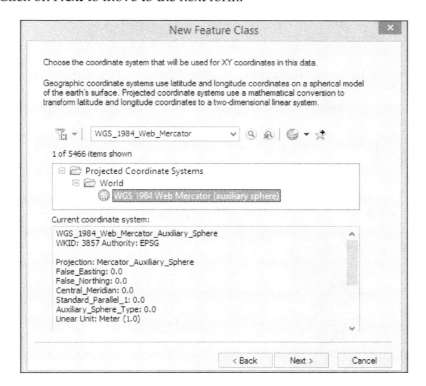

Next, we set the **XY Tolerance** value for the feature class. As you start adding features, you might want to add some features close to each other yet you don't want them to snap into the same position; in that case make this value smaller to get a higher accuracy for each feature position. However, sometimes you will need to add features on top of each other, making them overlap on purpose, especially if you have Z values. Too small a tolerance value might make it difficult to snap those features into a single location and might cause problems with shared boundaries.

9. As you can see, this value needs to be carefully planned, but for now, leave **XY Tolerance** to its default value, which is 0.001 meters, and click on **Next**.

 XY Tolerance is the minimum distance after which two features will snap together.

10. In the next form, we select the configuration keyword, choose the **Default Configuration Keyword**, and click on **Next**.

 The configuration keyword is a table space in which feature classes and tables are stored. Each configuration has certain properties such as the geometry type and file size that is shared by all objects in that keyword.

11. Finally, we add the fields for our feature class. Note that two fields are already added for you, OBJECTID, which is also the primary key, and a sequence number representing each feature uniquely in the feature class.

 The primary key is a column by which a record is uniquely identified in a table or a dataset.

The second field is SHAPE, which, if you remember, we added by specifying the geometry type. So, we need to add the two fields TOWER_ID and TOWER_RANGE_METERS. Click on an empty row in the **Field Name** column and add the following fields:

Name	Data Type
TOWER_ID	Text
TOWER_RANGE_METERS	Short integer

12. After adding all the fields, your dialog should look like the following screenshot. Click on **Finish** to create the feature class:

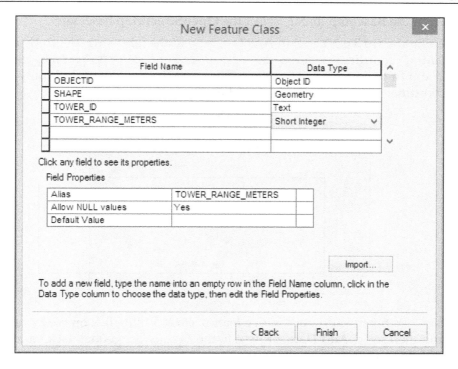

You just created your first feature class; if you take a look at the icon, you will see three small dots, which indicate that this feature class has point geometry.

Preparing the map document

Now that we have the geodatabase, we will need to populate it with towers, and for that we will require ArcMap. In this exercise, we will require some online base map to work with, which means we need an Internet connection to download the world base map. Follow these steps to prepare the document:

1. Open ArcMap.
2. Go to **File** | **Add Data**, and then click on **Add Basemap**.

3. From the **Basemap** dialog, select the **OpenStreetMap** basemap.

4. Using the **Go To XY** tool in ArcMap, type in the following longitude and latitude coordinates: Long: 2.326470; Lat: 48.843279. You may use the **Zoom In** tool to adjust your view. These are the Boulevard du Montparnasse coordinates, as illustrated in the following screenshot:

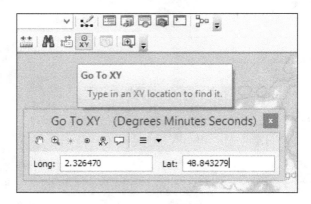

5. Now it's time to add our Towers feature class. Right-click on the **Layers** node and then select **Add Data**, browse to the telzaviba geodatabases and then add the Towers feature class.

6. Change the symbology of the Towers layer by using the Tower Short symbol. Refer to *Chapter 1, Getting Started with ArcGIS*, to learn how to change the symbology.

7. Double-click on the Towers layer and activate the **Labels** tab.

8. Check the **Show Labels** checkbox and then select the **Tower ID** field to display the tower ID for each tower on the map.

Now, it is time to add some features. Before you do so, make sure you close ArcCatalog and that you do not have any connections to your geodatabase. Now, to display the **Editor** toolbar, perform the following steps:

1. To display the Editor toolbar, right-click on an empty area in the menu and select the Editor toolbar to activate it.

2. On the Editor toolbar, point to **Editor** and then click on **Start Editing**.

3. Click on the **Create Features** button on the Editor toolbar and you will see the **Create Feature** window pop up on the right.

4. Click on **Cell Towers** and add the six towers with the following attributes along the boulevard, as shown in the following table:

Tower ID	Tower range
T01	60
T02	70
T03	60
T04	60
T05	40
T06	60

5. You can set the attributes for each feature you add by selecting that feature and clicking on the **Attribute** tool in the **Editor** toolbar. Add the towers, as illustrated in the following screenshot:

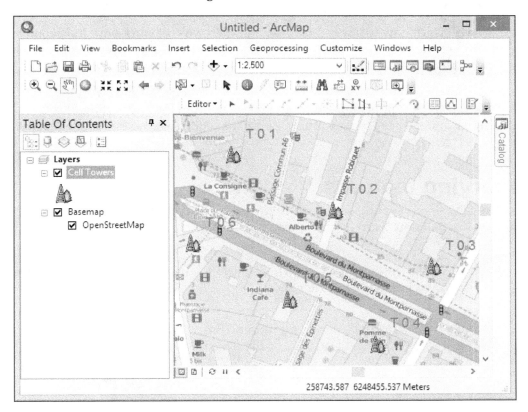

6. From the **Editor** toolbar, point to **Editor** and click on **Save Edits** if you wish to save your edits. Point to the **Editor** and then click on **Stop Editing**.

7. Close ArcMap.

You can find both the map document and the edited geodatabase under `B04847_02_Files\telzaviba\Data`.

Preparing the TelZaViBa add-in project

You can use the same steps we learned when we had created the Hello ArcGIS add-ins to create the TelZaViBA project. I have already prepared the empty project template; simply copy the project from the supporting files under `B04847_02_Files\telzaviba\Code`.

Displaying the range of each tower

Here we will learn how to query a feature, and then we will work with the geometry object and use the buffer operation to create a circle around the tower to represent the range based on a fixed radius.

In this section, we will start writing the actual ArcGIS code. To start working, copy the TelZaViBA empty add-in project template from the supporting files for this chapter under `B04847_02_Files\telzaviba\Code` to `C:\ArcGISByExample\telzaviba\Code`.

Querying features

Just like tables consist of records, feature classes consist of records that are called features. We will now learn how to query a feature when we have its object ID:

 The **ObjectID** of a feature is a primary key that uniquely identifies each feature in a feature class.

1. Open the TelZaViBa solution by running the `C:\ArcGISByExample\telzaviba\Code\TelZaViBA.sln` file. This will open Microsoft Visual Studio.

If you are using an older version of Visual Studio (prior to 2013), you can open the project directory from the `C:\ArcGISByExample\ telzaviba\Code\TelZaViBA\TelZaViBA.vbproj` folder.

2. Double-click to edit the `btnShowTowerRange.vb` class and add the following import statements before the class declaration:

```
Imports ESRI.ArcGIS.ArcMapUI
Imports ESRI.ArcGIS.Carto
Imports ESRI.ArcGIS.Geodatabase
Imports ESRI.ArcGIS.Geometry
Imports ESRI.ArcGIS.Display
```

3. If any of the above libraries failed to load, make sure to add the respective ArcGIS reference. Point to the **Project** menu and click on **Add Reference**. Type `Esri.ArcGIS` in the search box and add the missing reference to your project. You might need to follow this step if you are using an older version of ArcGIS prior to 10.3.

4. On the `OnClick` event, write the following code:

```
Dim pDocument As IMxDocument = My.ArcMap.Application.Document
Dim pTowerLayer As IFeatureLayer = pDocument.FocusMap.Layer(0)
Dim pFeature As IFeature = pTowerLayer.FeatureClass.GetFeature(1)

MsgBox("Feature with objectid " & pFeature.OID &
" - with class name " & pFeature.Class.AliasName &
" has been fetched")
```

Using the application handle we will get the current map document (pDocument), and then from the document we will get the first layer in the document assuming it is the one we are looking for (our Towers layer). Then from the Towers layer, we will access the underlying feature class and use the `getFeature` method to get the feature with object ID 1.

5. Build your project and make sure you don't have any errors.

6. From the supporting files in this chapter, copy the `B04847_02_Files\ telzaviba\Data` folder to `C:\ArcGISByExample\telzaviba\Data`. This will copy both the map and the data.

7. Open the `telzaviba.mxd` document under `C:\ArcGISByExample\ telzaviba\Data\`.

I have prepared 10.0, 10.1, and 10.2 map document copies depending on your ArcGIS version.

8. From the **Customize** menu, point to **Toolbars** and activate **Cell Tower Analysis Tool**. The toolbar has one button, which is the **Show Tower Range** button, as you can see in the following screenshot:

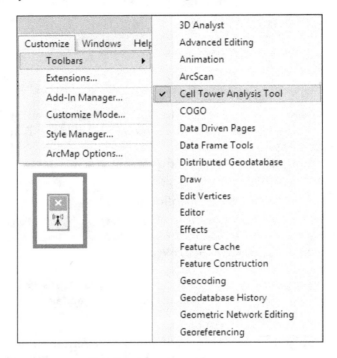

9. Click on the **Show Tower Range** button which has the Tower icon, you should see the following message:

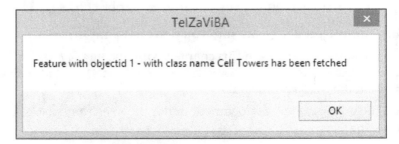

10. Close ArcMap and don't save any changes.

Now it is time to work with the geometry object and create a circle around the feature to represent the tower's signal range.

The topological operators and graphic elements

The topological operators allows us to work and manipulate the feature geometries and change, expand or shrink the shape. We will use the `Buffer` method to make the point tower feature into a circle and then we will draw this circle on the map. Follow these steps:

1. If you have closed it, open your TelZaViBa project again from `C:\ArcGISByExample\telzaviba\code\`.

2. Double-click to edit `btnShowTowerRange.vb` and add the following code in the `onClick` event:

   ```
   Dim pDocument As IMxDocument = My.ArcMap.Application.Document
   Dim pTowerLayer As IFeatureLayer = pDocument.FocusMap.Layer(0)
   Dim pFeature As IFeature = pTowerLayer.FeatureClass.GetFeature(1)

   Dim pTopo As ITopologicalOperator = pFeature.Shape
   Dim pTowerRange As IGeometry = pTopo.Buffer(100)

   MsgBox("Feature with objectid " & pFeature.OID &
   " - with class name " & pFeature.Class.AliasName &
    " has been fetched")
   ```

3. This has created a new polygon geometry of type `circle` with a radius of 100 meters. Now we need to draw that on the map using the ArcGIS graphics. Write the following code to do so:

   ```
   Dim pDocument As IMxDocument = My.ArcMap.Application.Document
   Dim pTowerLayer As IFeatureLayer = pDocument.FocusMap.Layer(0)
   Dim pFeature As IFeature = pTowerLayer.FeatureClass.GetFeature(1)

   Dim pTopo As ITopologicalOperator = pFeature.Shape
   Dim pTowerRange As IGeometry = pTopo.Buffer(100)

   Dim pElement As IElement = New PolygonElement
   pElement.Geometry = pTowerRange
   Dim pFillShapeElement As IFillShapeElement = pElement
   Dim pFillShapeSymbol As ISimpleFillSymbol = New SimpleFillSymbol
   pFillShapeSymbol.Style = esriSimpleFillStyle.esriSFSHorizontal
   pFillShapeElement.Symbol = pFillShapeSymbol
   ```

```
pDocument.ActiveView.GraphicsContainer.AddElement(pElement, 0)
    pDocument.ActiveView.Refresh()

MsgBox("Feature with objectid " & pFeature.OID &
" - with class name " & pFeature.Class.AliasName &
  " has been fetched")
```

4. Open `telzaviba.mxd` and click on the **Show Towers Range** button; you should see the circle surrounding our feature with object ID 1 with horizontal tiles, as shown in the following screenshot. Note that I used the measure tool to measure the radius of the created circle:

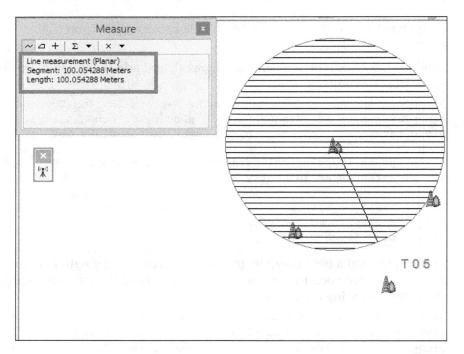

5. Close ArcMap and Visual Studio.

Drawing the tower range based on the attribute value

Finally, we hit the attributes for the feature and use it to redraw our tower ranges based on the tower attribute value range. We will apply the code that we wrote in the previous section, on all the towers. However, instead of using the hardcoded value of 100 meters, we will use the actual tower range.

Drawing the range attribute

We will use the range attribute to draw the signal buffer for one tower. Follow these steps:

1. Open your TelZaViBa project from C:\ArcGISByExample\telzaviba\ code\.

2. Change the following code in the onClick event for btnShowTowerRange.vb:

    ```
    ...
    Dim pDocument As IMxDocument = My.ArcMap.Application.Document
    Dim pTowerLayer As IFeatureLayer = pDocument.FocusMap.Layer(0)
    Dim pFeature As IFeature = pTowerLayer.FeatureClass.GetFeature(1)

    Dim pTopo As ITopologicalOperator = pFeature.Shape
    Dim range As Double = pFeature.Value(pFeature.Fields.
    FindField("TOWER_RANGE_METERS"))
    Dim pTowerRange As IGeometry = pTopo.Buffer(range)

    Dim pElement As IElement = New PolygonElement
    ...
    ```

3. Build your solution, open telzaviba.mxd, and run **Show Range Tower**. You should now get a different range since T01 has a 60 meter range.

4. Close ArcMap.

Drawing the range for all towers

So far we have managed to write code to draw the range for one tower only. What is missing is to be able to loop through all towers on the map and execute the same code on all of them. In order to do that, we need to learn about the feature cursor. The feature cursor allows us to traverse multiple features and process each one individually:

1. If you have closed your project, open it again from C:\ArcGISByExample\ telzaviba\code\.

2. Add the following code to your onClick event. Note that we have removed the message box code since we don't need it anymore:

    ```
    Dim pDocument As IMxDocument = My.ArcMap.Application.Document
    Dim pTowerLayer As IFeatureLayer = pDocument.FocusMap.Layer(0)

    Dim pFeature As IFeature

    Dim pFeatureCursor As IFeatureCursor =
      pTowerLayer.FeatureClass.Search(Nothing, False)
    ```

```
pFeature = pFeatureCursor.NextFeature

Do Until pFeature Is Nothing

Dim pTopo As ITopologicalOperator = pFeature.Shape
Dim range As Double = pFeature.Value(pFeature.Fields.
FindField("TOWER_RANGE_METERS"))
Dim pTowerRange As IGeometry = pTopo.Buffer(range)

Dim pElement As IElement = New PolygonElement
pElement.Geometry = pTowerRange
Dim pFillShapeElement As IFillShapeElement = pElement
Dim pFillShapeSymbol As ISimpleFillSymbol = New SimpleFillSymbol
pFillShapeSymbol.Style = esriSimpleFillStyle.esriSFSHorizontal
pFillShapeElement.Symbol = pFillShapeSymbol
pDocument.ActiveView.GraphicsContainer.AddElement(pElement, 0)
pDocument.ActiveView.Refresh()

pFeature = pFeatureCursor.NextFeature
Loop
```

3. Build and run your **Show Tower Range**, you should get something like the
    following screenshot:

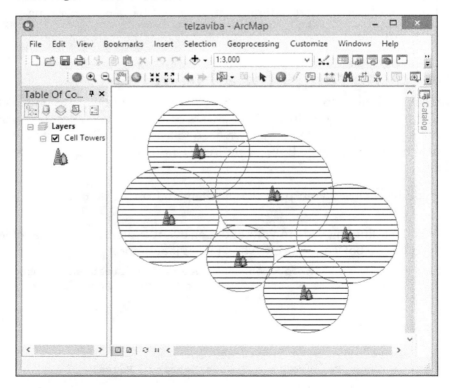

4. You can find the latest source code for TelZaViBa under the
    `B04847_02_Files\telzaviba\FinalCode` folder.

5. Close ArcMap and Visual Studio.

Summary

In this chapter, you learned about some new concepts in ArcGIS such as layers, feature class, features, and geometries. You also learned about graphic elements and symbols and how useful they proved as drawing tools on the map. We then introduced the topological operator and learned how we can use it to do spatial operations on geometries like what we did with the buffer. We used all these tools to draw the signal range for the towers on the map based on each tower's range value.

In the next chapter, you will learn about the proximity tools and you will then use them to estimate the signal strength. To do that, you will be introduced to some more add-ins controls.

3
Mapping Signal Strength

In the previous chapter, we prepared the TelZaViBA geodatabase and map document. We created the TelZaViBa Visual Studio project where we added the show tower range button. You also learned about the geodatabase and the feature classes, and some basic geometry operations. However, we only managed to show the range of the tower, and calculating the signal strength for a given point on the map is a different story. In this chapter, we will do some more geometrical operations on our towers to calculate the signal strength.

In this chapter, we will discuss the following topics:

- Adding a point to the map
- Finding the distance between two points
- Finding and highlighting the closest tower
- Displaying signal strength

Adding a point to the map

In this section, we will first discuss and explain the ArcGIS coordinate system. Then we will use a new add-in control to interact with the map. We will use the tool control to add a point to the map. This point, which is identified by two important parameters, x and y, represents a cell phone location.

The ArcGIS coordinate system

ArcGIS supports two different coordinate systems. The geographic coordinate system that uses decimal degrees (longitude and latitude) and the other one is the projected coordinate system that uses metric (x and y). There are just different ways to identify a location, and in this book we will be using the projected coordinate system.

To learn more about the ArcGIS coordinate system and the difference between the geographic and projected systems, refer to `http://bit.ly/b04748_coordinatesystems`.

To illustrate the coordinate system, follow these steps:

1. Open the `telzaviba.mxd` map document under the supporting files `C:\ArcGISByExample\telzaviba\Data\`. You can copy it from the supporting files for this chapter.

2. Bring up **Draw** by going to the **Customize** menu, point to **Toolbars**, and then click on **Draw**, as illustrated in the following screenshot. The **Draw** toolbar will allow us to add graphics to the map.

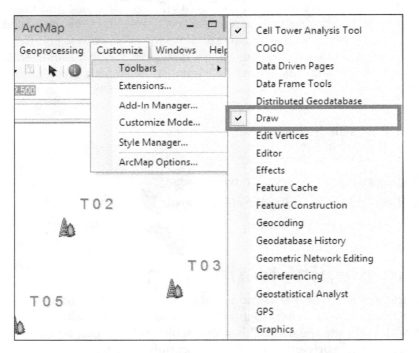

3. From the **Draw** toolbar, you can draw different graphics with multiple geometries. Point to the rectangle icon and choose the **Marker** point since we will be drawing points, as shown in the next screenshot:

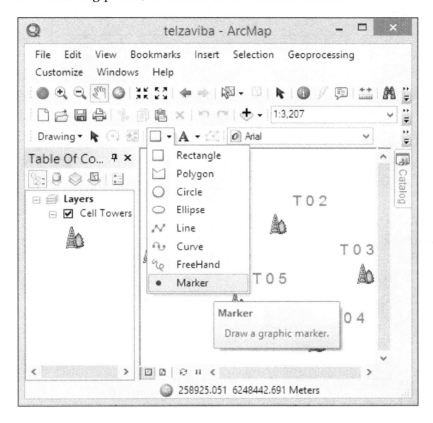

4. Hide the **Cell Towers** layer for now, to remove the clutter from the map, by unchecking the **Cell Towers** layer.

5. Use the **Marker** tool to add a point to the map; you will notice that a default green dot will be added to the map.

6. Double-click on the **Marker** point to open the graphics properties. Here you can change the color and properties of this graphic. Click on the **Location** tab and take note of the location of the point. This is the x and y coordinate for that particular marker, as shown in the following screenshot:

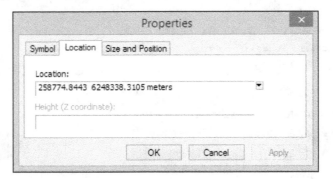

7. Now add five points, as illustrated in the next screenshot; I have manually added a rounded location to each point. Note how it looks like a regular x and y plane. The x axis increases from left to right while the y axis decreases from top to bottom.

Note how the three horizontal points have the same *y* coordinate 6248340, and their *x* values increases as you go from left to right. The other three vertical points have the *x* value of 258869 and note how the value of *y* decreases as you go from top to bottom.

8. Close ArcMap and choose not to save any changes.

Adding the find closest tower tool

The tool add-in allows interactions with the map. We have just shown a tool in the previous section where we added a point using the **Draw** toolbar. When you select a tool, you can click on the map and get the location. You can use this location to do so many things with it. In this section, we will add our own tool that displays the current location, and then we will add more functionality to it. Follow these steps to add the tool:

1. Open our TelZaViBa project; you can pick up from where we left off back in *Chapter 2*, *App 1 – the Cell Tower Analysis Tool*. If you are not sure, you can get the latest source code from this supporting files code under B04847_03_Files\telzaviba\StartCode and copy it to C:\ArcGISByExample\telzaviba\Code. Just make sure to delete any code in the folder before copying.

2. From the Visual Studio window, point to the **Project** menu and click on **Add Class**.

3. From the dialog, expand **Common Items | ArcGIS** and then click on **Desktop Add-ins**. Select **Add-in Component** and type tlFindClosestTower in the **Name** field. Click on **Add** to add the class to your project.

4. From the add-in wizard, select **Tool** from the **Add-in Component** panel.

5. Type `Find the Closest Tower to the Point` in **Caption, Tooltip,** and **Description**. Then type `TelZaViBa` in **Category**. Select the cell phone picture for the tool image; you can find the images under `B04847_03_Files\telzaviba\Icons`. Your form should look like the following screenshot:

6. Click on **Finish** to add the tool.

7. Now we need to add this tool to the cell tower analysis tool toolbar. To do that, double-click on the `Config.esriaddinx`, find the `<Toolbars>` XML tag, and add the following highlighted code:

```
<Toolbars>
    <Toolbar id="TelZaViBA_Cell_Tower_Analysis" caption="Cell
Tower Analysis Tool" showInitially="false">
        <Items>
          <Item refID="TelZaViBA_btnShowTowerRange" />
          <Item refID="TelZaViBA_tlFindClosestTower" />
        </Items>
    </Toolbar>
</Toolbars>
```

8. Open `tlFindClosestTower` by double-clicking it.

9. You should see a couple of empty methods called `New` and `OnUpdate`. What we are interested in is `onMouseUp`, that is the event which is fired when you click on the map and release your mouse button. The `onMouseUp` event is not there, so we have to add it manually. Write the following code to override the `onMouseUp` sub:

```
Protected Overrides Sub OnMouseUp(arg As ESRI.ArcGIS.Desktop.
AddIns.Tool.MouseEventArgs)

End Sub
```

Note that you can autogenerate this code by writing `Overrides OnMouseUp` and then select the method from the list shown in the following screenshot:

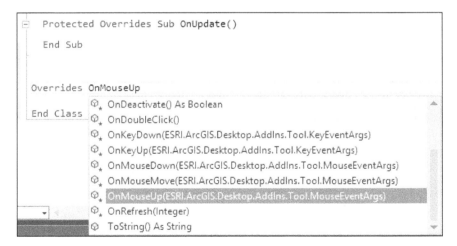

10. We want to display a message box with the current coordinates. Note that the `arg` parameter is the one carrying all this useful information. Write the following code to do that:

```
Protected Overrides Sub OnMouseUp(arg As ESRI.ArcGIS.Desktop.
AddIns.Tool.MouseEventArgs)

MsgBox("X: " & arg.X & "  ---  " & "Y: " & arg.Y)
End Sub
```

11. From the **Project** menu, select **Build** and run `telzaviba.mxd` to test our tool.

12. Note that we have a new icon besides our **Show Towers Range**. Click on it and click on any location on the map. You should see the message displaying the mouse coordinates, as shown in the following screenshot:

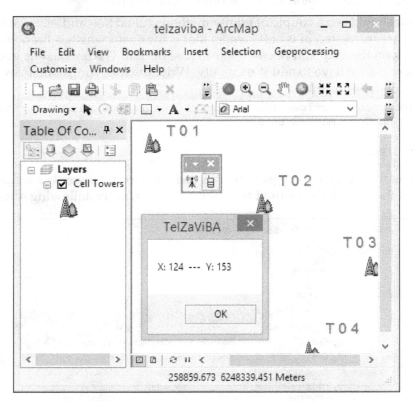

13. Notice how these coordinates are relatively small compared to the actual coordinates we got from the **Draw** tool. The reason is those coordinates are mouse coordinates relative to the window of ArcMap and not to the actual map coordinates. So how can one convert the mouse coordinates to map coordinates? We can do that with the help of the display transformation API in ArcObjects. Before we do that, we need to import a few libraries, as follows:

```
Imports ESRI.ArcGIS.ArcMapUI
Imports ESRI.ArcGIS.Geometry
Imports ESRI.ArcGIS.Carto
Imports ESRI.ArcGIS.Display
Imports ESRI.ArcGIS.Geodatabase

Public Class tlFindClosestTower
    Inherits ESRI.ArcGIS.Desktop.AddIns.Tool
...
```

14. Now add the following code in onMouseUp so that we can get a point geometry from the mouse coordinates:

```
Protected Overrides Sub OnMouseUp(arg As ESRI.ArcGIS.Desktop.
AddIns.Tool.MouseEventArgs)

MsgBox("X: " & arg.X & "  --- " & "Y: " & arg.Y)

Dim pDocument As IMxDocument = My.ArcMap.Application.Document
Dim pPoint As IPoint = pDocument.ActiveView.ScreenDisplay.
DisplayTransformation.ToMapPoint(arg.X, arg.Y)

MsgBox("Map X: " & pPoint.X & " ---- " & "Map Y: " & pPoint.Y)

End Sub
```

15. Build your solution and run telzaviba.mxd.

16. Activate your new tool and click on any location on the map; you should get two messages, the first is the mouse coordinates and the second one is the actual map coordinates, as shown in the following screenshot:

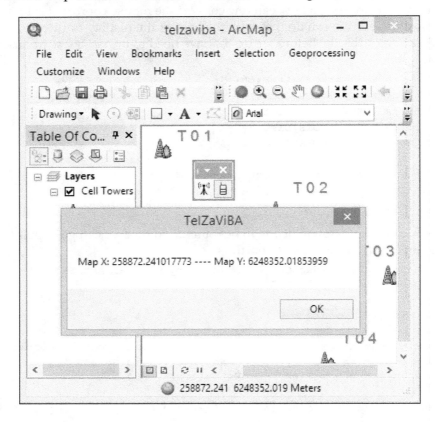

17. Close ArcMap and choose not to save any changes.

We are now ready to use this point geometry for further processing, graphic creation, and more.

Creating a marker point

In this section, we will write code to add a marker point on the map to display a picture of a cell phone. We can actually add a picture to the map by using the picture marker element. So far we have managed to create a tool and get the x and y coordinates of the mouse-click location. Then we learned how to create a point geometry using those coordinates.

Follow these steps to create a point on the map:

1. Open our TelZaViBa project from the `C:\ArcGISByExample\telzaviba\Code` if you have closed it.

2. We will now load the picture of the cell phone located in `C:\ArcGISByExample\telzaviba\Icons\Cell-phone.png` and add it to the map by creating a marker element and using the point we created in the previous section as the geometry:

```
Protected Overrides Sub OnMouseUp(arg As ESRI.ArcGIS.Desktop.
AddIns.Tool.MouseEventArgs)

Dim pDocument As IMxDocument = My.ArcMap.Application.Document
Dim pPoint As IPoint = pDocument.ActiveView.ScreenDisplay.
DisplayTransformation.ToMapPoint(arg.X, arg.Y)

Dim pElement As IElement = New MarkerElement
pElement.Geometry = pPoint
Dim pMarkerElement As IMarkerElement = pElement

Dim pPictureMarkerSymbol As IPictureMarkerSymbol = New
PictureMarkerSymbol
        pPictureMarkerSymbol.CreateMarkerSymbolFromFile(esriIPictu
reType.esriIPicturePNG, "C:\ArcGISByExample\telzaviba\Icons\Cell-
phone.png")
pPictureMarkerSymbol.Size = 20
pMarkerElement.Symbol = pPictureMarkerSymbol

pDocument.ActiveView.GraphicsContainer.AddElement(pElement, 0)
pDocument.ActiveView.Refresh()

End Sub
```

3. Build your solution and run `telzaviba.mxd`.

4. Click on your tool and click anywhere on the map; you should now start seeing cell phones being added to the map as you click on the map. This should look like the following screenshot:

5. To make things simple, let us make sure you can add a single cell phone only. Add the following line of code to clear all graphics before we add our new point to the map. This will ensure we have only one cell phone to work with at a time:

```
Protected Overrides Sub OnMouseUp(arg As ESRI.ArcGIS.Desktop.
AddIns.Tool.MouseEventArgs)

Dim pDocument As IMxDocument = My.ArcMap.Application.Document
Dim pPoint As IPoint = pDocument.ActiveView.ScreenDisplay.
DisplayTransformation.ToMapPoint(arg.X, arg.Y)
pDocument.ActiveView.GraphicsContainer.DeleteAllElements()

....
```

6. Close ArcMap and choose not to save any changes.

As you can see, we are slowly building an application using the add-in components. We are getting closer to satisfying our client's requirement.

Finding the distance between two points

In this section, we will learn about the proximity operators, which will help us to calculate the distance between two points. We will then use this to find the distance between the cell phone location and the tower. This will help us find the closest tower, which will be the connected tower.

> Proximity operator: This is an ArcObject method that returns the minimum distance between two geometric shapes. It works on points, lines, and polygon geometries.

Finding the distance between two towers

We will test our proximity operator on two of the towers, T01 and T02, in the TelZaViBa geodatabases to find the distance between them. We will get the T01 and T02 features using their corresponding object IDs and then use the **Shape** property to get the underlying point geometry under that. We do that by setting one geometry to the proximity operator object, then calling the `ReturnDistance` method, and passing the second geometry:

1. Open our TelZaViBa project from the `C:\ArcGISByExample\telzaviba\Code` if you have closed it.

2. Double-click on `btnShowTowerRange` to open its code and add the following code on the `onClick` event:

```
...
        pDocument.ActiveView.Refresh()


Dim t1 As IFeature = pTowerLayer.FeatureClass.GetFeature(1)
Dim t2 As IFeature = pTowerLayer.FeatureClass.GetFeature(2)

Dim pProximityOperator As IProximityOperator = t1.Shape
Dim distance As Double = pProximityOperator.ReturnDistance(t2.
Shape)

MsgBox("The distance between tower t01 and t02 is " & distance)

    End Sub
```

3. Build your solution and run `telzaviba.mxd`.

4. Click on the `ShowTowersRange` button; this will show the range for the towers, as we have seen in the previous chapter, and then it will display the distance between the message. The message says the distance is 102.9 meters between **T01** and **T02**. If we use the measuring tool, we can actually confirm that it is in fact true, as shown in the following screenshot:

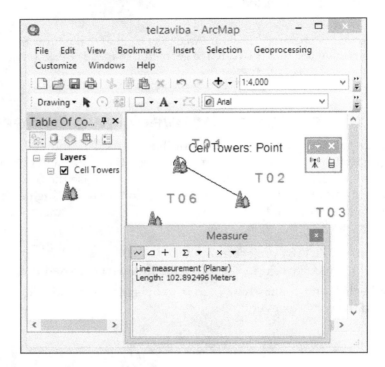

5. Close ArcMap and choose not to save any changes.

You can safely delete this code since we won't be requiring it for this method. However, we will be using it somewhere else.

Finding the distance between the point and a tower

Now that we know how to use the proximity tool, we will use it on our new tool, FindClosestTower. Follow these steps to find the distance between the cell phone and T01:

1. Open our TelZaViBa project from the `C:\ArcGISByExample\telzaviba` if you have closed it.

2. Double-click on the `tlFindClosestTower` class to edit it.

3. Add the following code to the `OnMouseUp` method after the refresh command:

```
pDocument.ActiveView.GraphicsContainer.AddElement(pElement, 0)
pDocument.ActiveView.Refresh()

Dim pProximityOperator As IProximityOperator = pPoint
Dim pTowerLayer As IFeatureLayer = pDocument.FocusMap.Layer(0)
Dim t1 As IFeature = pTowerLayer.FeatureClass.GetFeature(1)
Dim distance As Double = pProximityOperator.ReturnDistance(t1.
Shape)

MsgBox("Distance between the cell phone and tower T01 is " &
distance)

End Sub
```

4. Build your solution and run `telzaviba.mxd`.

5. Activate the **Find Closest Tower** tool and then click on the map, and compare the distance you get from the message with the distance from the ArcMap measure tool. This is illustrated in the following screenshot:

6. Close ArcMap and choose not to save any changes.

Finding and highlighting the closest tower

We know how to add a point on the map and we also know how to find the distance between two points. What remains is finding the closest tower to the cell phone we are adding on the map. To do that, we need to find the distance between the point we are adding and all the towers. Compare all the distances and the shortest distance is the closest to the point. We will then get the tower name and display it as a message. Later, we will learn how to use the display object to highlight and flash that tower.

Finding the closest tower

To find the closest tower, we need to loop through all towers and run the same proximity code. While looping, we will take note of the shortest distance and save the corresponding ObjectID. If we found a shorter distance, we will update ObjectID and that will be our closest tower:

1. Open our TelZaViBa project from the C:\ArcGISByExample\telzaviba\Code if you have closed it.

2. Double-click on t1FindClosestTower to edit it.

3. Delete all code after the refresh command and replace it with the following code:

```
pDocument.ActiveView.GraphicsContainer.AddElement(pElement, 0)

pDocument.ActiveView.Refresh()


Dim pTowerLayer As IFeatureLayer = pDocument.FocusMap.Layer(0)
Dim pFeatureCursor As IFeatureCursor = pTowerLayer.FeatureClass.
Search(Nothing, False)
Dim pFeature As IFeature = pFeatureCursor.NextFeature

Dim pProximityOperator As IProximityOperator = pPoint
Dim closestTowerOID As Long
Dim shortestdistance As Double = Double.MaxValue
```

```
Do Until pFeature Is Nothing

Dim distance As Double = pProximityOperator.
ReturnDistance(pFeature.Shape)
If distance < shortestdistance Then
    shortestdistance = distance
    closestTowerOID = pFeature.OID
End If

    pFeature = pFeatureCursor.NextFeature
Loop

MsgBox("Closest Tower is " & closestTowerOID & " with a distance
of " & shortestdistance)
```

4. Showing the object ID is not interesting, we need `TowerID`, and to get that, we need to tab into the feature and, therefore, its attributes. Just before the message box, add the following code to get the feature and `TowerID`. We will also round the distance to the nearest integer:

```
Dim distance As Double = pProximityOperator.
ReturnDistance(pFeature.Shape)
If distance < shortestdistance Then
    shortestdistance = distance
    closestTowerOID = pFeature.OID
End If

    pFeature = pFeatureCursor.NextFeature
Loop

Dim pClosestTower As IFeature = pTowerLayer.FeatureClass.
GetFeature(closestTowerOID)

MsgBox("Closest Tower is " & pClosestTower.Value(pClosestTower.
Fields.FindField("TOWER_ID")) & " with a distance of " & Math.
Round(shortestdistance) & " Meters")
```

5. Build your solution and run `telzaviba.mxd`.

6. Use your **Find Closest Tower** tool and add a cell phone near tower **T03**, for instance. You should see that our tool can now find the closest tower, which is basically the tower the cell phone will be connected to. Take a look at the following screenshot:

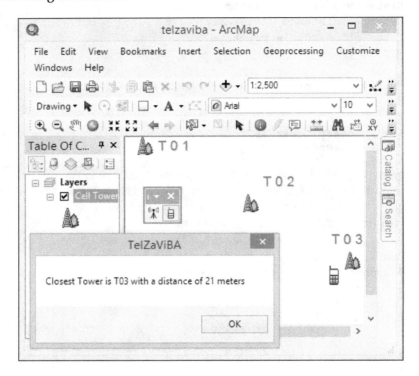

7. Try your tool with the rest of the towers and you will see how fascinatingly it works.

8. Close ArcMap and choose not to save any changes.

Highlighting the closest tower

Our tool is becoming rich as we start adding more functionalities to it. Now it is time to get rid of the annoying message box and replace it with something more user and map friendly. We will use the concept of flashing, or highlighting, where we will let the tower blink to identify its location.

Highlighting an object on the map is actually a little trick that is done on the screen as opposed to the actual map. We can do that using the display object in ArcObjects. Follow these steps to highlight the closest tower we have just found:

1. Open our TelZaViBa project from the
   `C:\ArcGISByExample\telzaviba\Code` if you have closed it.

2. Double-click on `tlFindClosestTower` to edit it.

3. Remove the message box and add the following code instead:

```
Dim pClosestTower As IFeature = pTowerLayer.FeatureClass.
GetFeature(closestTowerOID)

Dim pScreenDisplay As IScreenDisplay = pDocument.ActiveView.
ScreenDisplay
pScreenDisplay.StartDrawing(pScreenDisplay.hDC, ESRI.ArcGIS.
Display.esriScreenCache.esriNoScreenCache)

Dim pMarkerSymbol As IMarkerSymbol = New SimpleMarkerSymbol
Dim pColor As IRgbColor = New RgbColor
pColor.RGB = RGB(255, 0, 0)
pMarkerSymbol.Color = pColor
pMarkerSymbol.Size = 10
Dim pSymbol As ISymbol = pMarkerSymbol
pSymbol.ROP2 = esriRasterOpCode.esriROPNotXOrPen

pScreenDisplay.SetSymbol(pSymbol)

pScreenDisplay.DrawPoint(pClosestTower.Shape)
Threading.Thread.Sleep(200)
pScreenDisplay.DrawPoint(pClosestTower.Shape)
Threading.Thread.Sleep(200)
pScreenDisplay.DrawPoint(pClosestTower.Shape)
Threading.Thread.Sleep(200)
pScreenDisplay.DrawPoint(pClosestTower.Shape)

pScreenDisplay.FinishDrawing()
```

4. Unfortunately, I cannot show you the tower flashing in this book, however, you can visit my YouTube channel, where I have posted a video and explained flashing in detail, at `http://bit.ly/b04748_flashgeo`.

5. Build your solution and run `telzaviba.mxd`.

6. Use your new **Find Closest Cell Tower** tool and you should start to see that the closest tower flashes when you add the cell phone near it.

7. Close ArcMap and choose not to save any changes.

Displaying the signal strength

Finally, all that we have done during this chapter had one purpose: to display the strength of the signal. We managed to add the cell phone on the map, find the closest tower, and even highlight that tower, but now we need to do some math to calculate the signal. To calculate the signal in terms of percentage, we should establish some rules. If you are sitting just right by the tower, you should get 100 percent signal, and if you are sitting exactly on the edge of the range of the tower, you should have 0 percent signal. So we can deduce that the formula for calculating the signal is as follows:

```
S = (R - d)/R * 100
```

In the preceding formula, S is the signal strength in percentage, R is the range of the connected tower, and d is the distance between the cell phone and the tower. Of course, if we get a minus signal, that means we are out of range of the signal.

 Disclaimer: This example is from my imagination; I'm pretty sure that calculating the signal is much more complicated than that. In this book, we used such a formula for the sake of explanation.

1. Open our TelZaViBa project from the `C:\ArcGISByExample\telzaviba` if you have closed it.
2. Double-click on `tlFindClosestTower` to edit it.
3. On the `onMouseUp` event, add the following code at the end of the method just after the flashing code. This code will add the signal strength and the tower ID to the map. In the following code, we will use the text element, which allows you to add text to the map:

```
pScreenDisplay.FinishDrawing()

Dim t As String = pClosestTower.Value(pClosestTower.Fields.
FindField("TOWER_ID"))
Dim r As Double = pClosestTower.Value(pClosestTower.Fields.
FindField("TOWER_RANGE_METERS"))
Dim s As Double = Math.Round((r - shortestdistance) / r * 100)


Dim pTextPoint As IPoint = pPoint
pTextPoint.Y = pTextPoint.Y - 20
Dim pTextElement As ITextElement = New TextElement
pTextElement.Text = t & "(" & s & "%)"
```

```
Dim pTextSymbol As ITextSymbol = New TextSymbol
pTextSymbol.Size = 5

Dim pTheTextElement As IElement = pTextElement
pTheTextElement.Geometry = pTextPoint
pDocument.ActiveView.GraphicsContainer.AddElement(pTheTextElement,
0)
pDocument.ActiveView.Refresh()
```

4. Note how we created a new point for our text element just to move it down a bit, 20 meters, to avoid the text overlapping with the cell phone icon. Build your solution and run `telzaviba.mxd`.

5. Run the **Closest Tower** tool. After the flashing of the closest tower, you should see a text element added with the signal strength and the tower ID, as illustrated in the following screenshot:

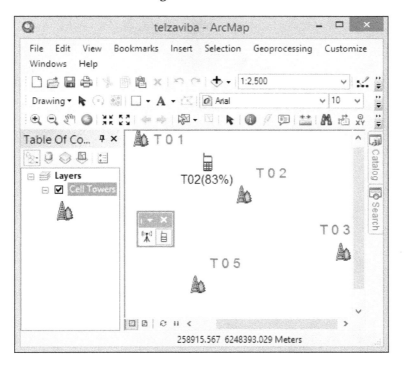

6. Close ArcMap and the Visual Studio.

Finally, we have managed to show the signal strength based on so many factors. You can, in fact, remove the code that deletes all elements and test out multiple cell phones if you would like, it will work fine. You can find the latest code under `B04847_03_Files\telzaviba\FinalCode`.

Summary

In this chapter, you injected our TelZaViBa tool with more functionality. You learned about the ArcGIS coordinates system and how to interact with the map by adding a tool. You then used this tool to add a point to simulate the location of a cell phone. Using the proximity operators, you have used this location to find the closest tower to the cell phone, which is the one the cell phone will be connected to. You were introduced to the display object, which was used to highlight and flash the connected tower. After completing all these tasks, you were ready to finally achieve the chapter aim by displaying the signal strength of that particular cell phone.

In the next chapter, you will complete the TelZaViBa tool by doing the real-time maneuvering; you will learn to read GPS coordinates from a text file, convert them into map points, and simulate a moving cell phone that constantly updates its signal.

4
Real-time Maneuvering

In the previous chapter, we managed to develop one of the important pieces of our TelZaViBa solution, that is, to display the signal strength for a connected cell phone. The exercise was broken into pieces and each was solved independently. You learned how to add a cell phone as a point to the map and use proximity operators to find the closest tower to the cell phone. Finally, we used the connected tower range to calculate the cell phone signal strength and display it on the map.

In the next pages, we want to elevate this example by simulating real people with cell phones walking on the Boulevard du Montparnasse. We will read external GPS coordinates and load them into our application to simulate someone walking on the boulevard.

In this chapter, we will discuss the following topics:

- Adding real-time cell phone simulator button add-ins
- Reading and mapping external GPS point coordinates
- Enabling signal maneuvering and highlighting the active tower

Adding real-time cell phone simulator button add-ins

We have seen add-on buttons before. In this section, we will add a button class where all our code will be written. This button, which will have a GPS icon, will be the sole work in this chapter. Eventually, we want to click on this button and display a form to ask the user to select the GPS file, which has a list of previously recorded tracks of coordinates. Then we will parse this file, process each record individually, and display that particular point on the map. We will create a timer to fetch each coordinate every second so that we will be able to actually see the cell phone signal strength being refreshed. We will also make this refresh interval a configurable value so that we can change it and see how fast this can progress. Follow these steps to add the button:

1. Open our TelZaViBa project, you can pick up from where we left off back in *Chapter 3*, *Mapping Signal Strength*. If you are not sure, you can get the latest source code from the supporting files code under `B04847_04_codes\telzaviba\StartCode\Telzaviba` and copy it to `C:\ArcGISByExample\telzaviba\Code`. Just make sure to delete any code in the folder before copying.

2. From the Visual Studio window, point to the **Project** menu and click on **Add Class**.

3. From the dialog, expand **Common Items | ArcGIS**, and then click on **Desktop Add-ins**. Select **Add-in Component** and type `btnRealTime` in the **Name** field. Click on **Add** to add the class to your project.

4. From the Add-in wizard, select **Button** from the **Add-in Component** panel.

5. Type `Simulate Real-Time Cell Tower Maneuvering` in **Caption**, **Tooltip**, and **Description**. Type `TelZaViBa` in **Category**. Select the GPS picture for the button image; you can find the images under `B04847_04_code\telzaviba\Icons`. Your form should look like the following screenshot:

6. Click on **Finish** to add the button class.

7. Now we need to add this button to the **Cell Tower Analysis Tool** toolbar. To do that, double-click on Config.esriaddinx, find the <Toolbars> XML tag, and add the following highlighted code:

```
<Toolbars>
    <Toolbar id="TelZaViBA_Cell_Tower_Analysis" caption="Cell
Tower Analysis Tool" showInitially="false">
        <Items>
          <Item refID="TelZaViBA_btnShowTowerRange" />
          <Item refID="TelZaViBA_tlFindClosestTower" />
          <Item refID="TelZaViBA_btnRealTime" />
        </Items>
    </Toolbar>
</Toolbars>
```

8. Open `btnRealTime` by double-clicking on it.

9. Add the following libraries at the top of the class for later use:

    ```
    Imports ESRI.ArcGIS.ArcMapUI
    Imports ESRI.ArcGIS.Geometry
    Imports ESRI.ArcGIS.Carto
    Imports ESRI.ArcGIS.Display
    Imports ESRI.ArcGIS.Geodatabase

    Public Class btnRealTime
        Inherits ESRI.ArcGIS.Desktop.AddIns.Button
    ...
    ```

10. Write the following code in the `onClick` event; this will display the ArcMap title document:

    ```
    Protected Overrides Sub OnClick()

        MsgBox(My.ArcMap.Application.Caption)

    End Sub
    ```

11. From the **Project** menu, select **Build** and run `telzaviba.mxd` to test our tool. If you are starting fresh from this chapter, you can copy `telzaviba.mxd` from `B04847_04_Files\telzaviba\Data` to `C:\ArcGISByExample\telzaviba\Data` and you can run `telzaviba.mxd` from `C:\ArcGISByExample\telzaviba\Data`.

12. Note that we have a new icon on our toolbar. Click on it and make sure that this code executes. You should see something similar to the following screenshot:

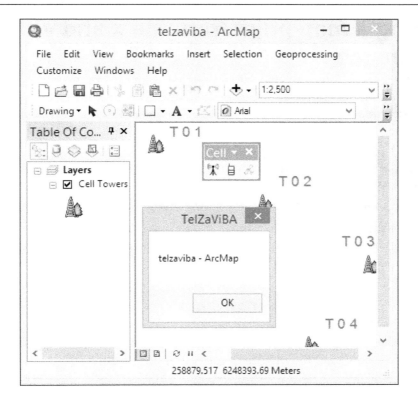

13. Close ArcMap and choose not to save any changes.

We are now ready to start working on our real-time maneuvering scenario.

Creating a map point from the x and y coordinates

In this section, we will we take the following coordinate (258825.9388, 6248364.5863), which I prepared beforehand, and convert it into a map point, and then we will add a point to the map as we have done in the previous chapter.

Follow these steps to create the point:

1. Open our TelZaViBa project from C:\ArcGISByExample\telzaviba\Code if you have closed it.

2. Open btnRealTime by double-clicking on it.

3. Write the following code in the onClick event to create a new point geometry. You can delete the message box that we previously added:

```
Protected Overrides Sub OnClick()

    Dim pPoint As IPoint = New Point
    pPoint.PutCoords(258825.9388, 6248364.5863)

End Sub
```

4. We will now load the picture of the cell phone, located in C:\ArcGISByExample\telzaviba\Icons\Cell-phone.png, as a symbol and add it to the map by creating a marker element and using the point we created in the previous section as the geometry. PictureMarkerSymbol allows us to load a picture and add it to an element. Write the following code to add a picture to the map:

```
Protected Overrides Sub OnClick()

    Dim pPoint As IPoint = New Point
    pPoint.PutCoords(258825.9388, 6248364.5863)


    Dim pDocument As IMxDocument = My.ArcMap.Application.Document
    Dim pElement As IElement = New MarkerElement
    pElement.Geometry = pPoint
    Dim pMarkerElement As IMarkerElement = pElement

    Dim pPictureMarkerSymbol As IPictureMarkerSymbol = New PictureMarkerSymbol
    pPictureMarkerSymbol.CreateMarkerSymbolFromFile(esriIPictureType.esriIPicturePNG, "C:\ArcGISByExample\telzaviba\Icons\Cell-phone.png")
```

```
pPictureMarkerSymbol.Size = 20
pMarkerElement.Symbol = pPictureMarkerSymbol

pDocument.ActiveView.GraphicsContainer.
AddElement(pElement, 0)
pDocument.ActiveView.Refresh()

End Sub
```

5. Build your solution and run `telzaviba.mxd`.

6. Click on your **Real Time Maneuvering** button and you should see that the cell phone icon has been added to a specific location on the map, as shown in following screenshot:

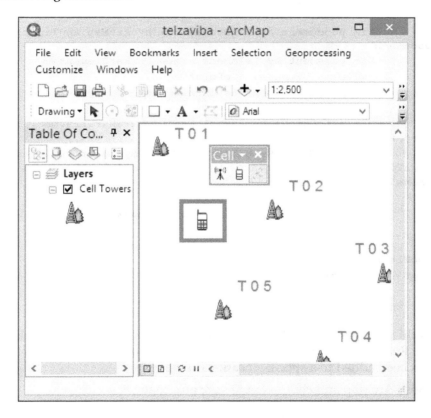

7. Close ArcMap and choose not to save any changes.

8. We need to organize and refactor our code for more clarity and readability. We will first create a new method `RunManeuver` that accepts a parameter of type `IPoint`. Then we will move our code to that method and call the method, as illustrated in the following, new code:

```
Protected Overrides Sub OnClick()

    RunManeuver(258825.9388, 6248364.5863)

End Sub

Public Sub RunManeuver(X As Double, Y As Double)
    Dim pPoint As IPoint = New Point
    pPoint.PutCoords(X, Y)

    Dim pDocument As IMxDocument = My.ArcMap.Application.
Document
    Dim pElement As IElement = New MarkerElement
    pElement.Geometry = pPoint
    Dim pMarkerElement As IMarkerElement = pElement

    Dim pPictureMarkerSymbol As IPictureMarkerSymbol = New
PictureMarkerSymbol
    pPictureMarkerSymbol.CreateMarkerSymbolFromFile(esriIPic
tureType.esriIPicturePNG, "C:\ArcGISByExample\telzaviba\s\Cell-
phone.png")
    pPictureMarkerSymbol.Size = 20
    pMarkerElement.Symbol = pPictureMarkerSymbol

    pDocument.ActiveView.GraphicsContainer.
AddElement(pElement, 0)
    pDocument.ActiveView.Refresh()

End Sub
```

9. Build your solution and make sure your new code works fine by clicking on the **Real Time Maneuvering** button. You should see that the cell phone icon has been added to a specific location on the map.

10. Close ArcMap and choose not to save any changes.

The `RunManeuver` method we just added will make our work much easier for the next topic.

Reading and mapping external GPS point coordinates

Here we will discuss how to read GPS coordinates from an external file and then load them on the map. The GPS log coordinates have been previously prepared for this example. We have multiple GPS file logs that we can pick from. For this exercise, we will use `C:\ArcGISByExample\telzaviba\Data\GPS_LOG_straight.txt`.

Reading external GPS coordinates

We will read an external file with GPS coordinates under
`C:\ArcGISByExample\telzaviba\Data\GPS_LOG_straight.txt`.
The file has pairs of *x* and *y* records, as shown in the following table:

x	y
258831.891978741	6248372.52378893
258961.538071367	6248277.93505809
258952.939095835	6248281.90381602
258944.340120304	6248287.19549327
258936.402604429	6248290.50279156
258925.819249929	6248295.79446881
258914.574435773	6248301.08614605
258901.345242648	6248309.68512159

Follow these steps to read the GPS points from the file:

1. Open our TelZaViBa project from `C:\ArcGISByExample\telzaviba\Code` if you have closed it.

2. Double-click on `btnRealTime` to open its code and add the following code on the `onClick` event. Delete any code that we previously wrote in there. This will help us read each line in the file:

```
Protected Overrides Sub OnClick()


        Dim gpsFile As String =
"C:\ArcGISByExample\telzaviba\Data\GPS_LOG_straight.txt"
        Dim eLines As IEnumerable(Of String) = IO.File.
ReadLines(gpsFile)
```

```
For Each sLine As String In eLines

Next
```

```
End Sub
```

3. Now each line consists of two values separated by a comma, and in order to get each of them, we need to use the split command as follows. The first part is *x* while the second part is *y*. For simplicity, we will display these values as a message box:

```
Protected Overrides Sub OnClick()

Dim gpsFile As String = "C:\ArcGISByExample\telzaviba\Data\GPS_
LOG_straight.txt"
    Dim eLines As IEnumerable(Of String) = IO.File.
ReadLines(gpsFile)
        For Each sLine As String In eLines
            Dim x As Double = sLine.Split(",")(0)
            Dim y As Double = sLine.Split(",")(1)
            Msgbox "X :" & X & " - " & "Y : " & Y
        Next

    End Sub
```

4. Build and run your `telzaviba.mxd` file.

5. Click on the **Simulate RealTime Cell Tower Maneuvering** button; you should see a fleet of messages displaying the *x* and *y* values of each record in the `GPS_LOG_straight.txt` file. Drawing all of them, as we will see in the next section, will show the path of this cell phone.

6. Close ArcMap and choose not to save any changes.

Mapping GPS coordinates

Now that we have learned how to read GPS coordinates, we need to add them to the map. Follow these steps to do so. We will require an Internet connection for this step:

1. Open our TelZaViBa project from `C:\ArcGISByExample\telzaviba\Code` if you have closed it.

2. Double-click on `btnRealTime` to open its code.

3. To map these GPS values, we only need to call `RunManeuver` and pass the *x* and *y* values to it so that it does its work. Add the following code and replace the message box:

```
        Protected Overrides Sub OnClick()

Dim gpsFile As String = "C:\ArcGISByExample\telzaviba\Data\GPS_
LOG_straight.txt"
        Dim eLines As IEnumerable(Of String) =
IO.File.ReadLines(gpsFile)
        For Each sLine As String In eLines
           Dim x As Double = sLine.Split(",")(0)
           Dim y As Double = sLine.Split(",")(1)
RunManeuver(x, y)
        Next

        End Sub
```

4. Build your solution and run `telzaviba_withbasemap_require_internet.mxd`; alternatively, you can run `telzaviba.mxd` and add your own basemap point to the file, then add the data, and click on **Add BaseMap**.

 An Internet connection is required only to load the basemap; if you do not have an access, you can open `telzaviba.mxd` and your code will work fine. The only thing you won't be able to see is the street background data.

5. Click on the **Simulate RealTime Cell Tower Maneuvering** button; you should see that the GPS points track are all drawn on the map. After loading GPS_LOG_straight.txt, your map should look like the following screenshot:

6. Close ArcMap and choose not to save any changes.

We still have some work to do; we need to display the signal strength for each location point.

Enabling signal maneuvering and highlighting the active tower

Now that we have learned how to read and map the coordinates, we need to add the maneuver by fetching each GPS point in a specific interval instead of loading all of them to the map. This will allow us to see a real-time view of the actual path.

Loading the GPS file

To start simulating real-time maneuvering, we should first read the coordinates from the file into memory. To do that, follows these steps:

1. Open our TelZaViBa project from `C:\ArcGISByExample\telzaviba\Code` if you have closed it.

2. Double-click on `btnRealTime` to open its code.

3. First we will need to add a Windows Form to ask the user to select the GPS file instead of hardcoding it. To add a Windows Form, point to **Project** menu and then click on **Add Windows Form**. Then, in the **Add New Item** dialog, type `frmRealTime.vb` in the **Name** text, as shown in the following screenshot:

4. Click on **Add** to add the form to the project and double-click on frmRealTime to open it.

5. We will need a timer control to start reading each line. The timer object will have a method that will be called every 3 seconds (3000 milliseconds). We will set that to be configurable at a later stage. From the **ToolBox** panel, drag the **Timer** control and drop it on the form.

6. Now we need to add the OpenFileDialog control. This control will help us browse and look for a GPS file coordinate in order to load it. From the **ToolBox** panel, drag the OpenFileDialog control and drop it on the form.

7. Similarly, add a **Button** control from **Toolbox**. Click on **Button** and change its **Text** property to Load GPS File. Your new form should look like the following screenshot:

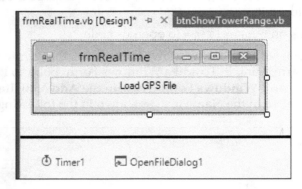

8. Now it is time to write some code in the form. Double-click on the **Load GPS File** button; this will add a new method for us to write the code in:

9. Write the following code in the Button1_Click method to open the file dialog:

```
Private Sub Button1_Click(sender As Object, e As EventArgs)
Handles Button1.Click
        OpenFileDialog1.ShowDialog()

        Dim gpsFile As String = OpenFileDialog1.FileName
        Dim eLines As IEnumerable(Of String) = IO.File.
ReadLines(gpsFile)
```

```
For Each sLine As String In eLines
    Dim x As Double = sLine.Split(",")(0)
    Dim y As Double = sLine.Split(",")(1)

    RunManeuver(x, y)
Next

End Sub
```

10. You will get an error on the `RunManeuver` line; to fix that, copy the `RunManeuver` method from `btnRealTime` to `frmRealTime`.

11. Import the following libraries at the beginning of the form:

```
Imports ESRI.ArcGIS.ArcMapUI
Imports ESRI.ArcGIS.Geometry
Imports ESRI.ArcGIS.Carto
Imports ESRI.ArcGIS.Display
Imports ESRI.ArcGIS.Geodatabase

Public Class frmRealTime
...
```

12. Double-click on the `btnRealTime` class to edit it.

13. Delete all the code on `onClick` and instead add the following code. This will show `frmRealTime`:

```
Protected Overrides Sub OnClick()

    Dim pRealTimeForm As New frmRealTime
    pRealTimeForm.Show()

End Sub
```

14. Build your solution and run `telzaviba_withbasemap_require_internet.mxd`; alternatively, you can run `telzaviba.mxd` and add your own basemap point to file, then add the data, and click on **Add BaseMap**.

15. Click on the **Simulate RealTime Cell Tower Maneuvering** button, which will show our new form.

16. Click on **Load GPS File**, which will show the open file dialog. Use the dialog to browse to our GPS file in `C:\ArcGISByExample\telzaviba\Data\GPS_LOG_straight.txt`.

17. You should see the GPS track being drawn on the map, as illustrated in the following screenshot:

18. Close ArcMap and don't save any changes.

Enabling maneuvering with a timer

Now that we have loaded the GPS file into memory, it is time to automate this process so that a timer will periodically read a coordinate at each specific interval. Follow these steps to do so:

1. If necessary, open your Visual Studio TelZaViBa source code.

2. Now we need to start writing the code to load our coordinates. The plan is to load the entire GPS file into memory and then loop on list by line. Each time the timer ticks, we will increment the line by one. This will move the `RunManeuver` method to the `Timer1_tick` method.

3. Double-click on `frmRealTime` to open the form designer.

4. Double-click on **Timer1** to open its underlying code. This will add a new method called **Timer1_Tick**.

5. Add the two modular variables, `_lines` for storing all the lines and `_currentline` to point to the current line:

```
Private _lines As New List(Of String)
Private _currentline As Integer = 0

Private Sub Timer1_Tick(sender As Object, e As EventArgs)
Handles Timer1.Tick

End Sub
```

6. Now we need to process every line with each timer tick. We will call the `RunManoeuve` method and increment the line after each load. Add the following code:

```
Private _lines As New List(Of String)
Private _currentline As Integer = 0

Private Sub Timer1_Tick(sender As Object, e As EventArgs)
Handles Timer1.Tick
     If _currentline >= _lines.Count Then Return

     Dim sLine As String = _lines(_currentline)
     Dim x As Double = sLine.Split(",")(0)
     Dim y As Double = sLine.Split(",")(1)
RunManeuver(x, y)
     _currentline = _currentline + 1
   End Sub
```

7. Go back to the `Button1_Click` method and modify the code to enable the timer, set the interval to tick the timer every 3 seconds (3000 milliseconds), and populate all the lines in the `_lines` list:

```
Private Sub Button1_Click(sender As Object, e As EventArgs)
Handles Button1.Click
     OpenFileDialog1.ShowDialog()

     Dim gpsFile As String = OpenFileDialog1.FileName
     Dim eLines As IEnumerable(Of String) = IO.File.
ReadLines(gpsFile)

     _lines = New List(Of String)
     _currentline = 0
```

```
           For Each sLine As String In eLines

                _lines.Add(sLine)
           Next

           Timer1.Enabled = True
           Timer1.Interval = 3000

      End Subs
```

8. To make things real-time, we need to clear all graphics before adding the new ones. Add the following line to delete all elements:

```
Public Sub RunManeuver(X As Double, Y As Double)

          Dim pPoint As IPoint = New Point
          pPoint.PutCoords(X, Y)

          Dim pDocument As IMxDocument = My.ArcMap.Application.
Document
          pDocument.ActiveView.GraphicsContainer.
DeleteAllElements()
```

9. Build your solution and run `telzaviba_withbasemap_require_internet. mxd`; alternatively, you can run `telzaviba.mxd` and add your own basemap point to the file, then add the data, and click on **Add BaseMap**.

10. Click on the **Simulate RealTime Cell Tower Maneuvering** button.

11. From the `frmRealTime` form, click on **Load GPS File** and then select the `C:\ArcGISByExample\telzaviba\Data\GPS_LOG_straight.txt` file.

12. Your code should work now and you should start seeing the cell phone move along the boulevard every 3 seconds.

13. Close ArcMap and choose not to save any changes.

14. Let us make this interval configurable; double-click on your `frmRealtime` form to open the **Form Designer**.

15. From **Toolbox**, drag the **TextBox** and **Label** controls onto the form.

16. Select your new label, which is named `Label1`, and in the **Text** property of **Label**, type `Refresh Interval`.

17. Select your new **TextBox**, which is named `TextBox1`, and in the **Text** property, type 3 to set the default refresh to 3 seconds.

18. Now we need to configure the code to read from our refresh text value. Double-click on **Load GPS File** to open the `Button1_Click` method. Change `Timer1.Interval` to the following code. We had to multiply by 1,000 in order to convert seconds to milliseconds:

```
Timer1.Interval = TextBox1.Text * 1000
```

19. Build your solution and run the `telzaviba_withbasemap_require_internet.mxd` document.

20. Click on the **Simulate RealTime Cell Tower Maneuvering** button.

21. From the `frmRealTime` form, type 1 in the **Refresh Interval** textbox to speed up the refresh interval. Click on **Load GPS File** and then select the `C:\ArcGISByExample\telzaviba\Data\GPS_LOG_straight.txt` file.

22. You should notice that the cell phone refreshes with a new coordinate every 1 second.

 You can even go less than a second to make the refresh even faster, try changing the refresh value to 0.5 seconds.

23. Close ArcMap and choose not to save any changes.

We have now completed the real-time maneuvering and we are almost done. Experiment with loading different GPS files such as `GPS_LOG_turn.txt` and `GPS_LOG_zigzag.txt`, and see what you get.

Highlighting the selected tower and displaying signal strength

We have already prepared and done this in *Chapter 3, Mapping Signal Strength,* and all we have to do in this section is to add that piece of code to our logic in the form. This code will display the signal strength and also it will highlight the connected tower. Follow these steps to do so:

1. Open our TelZaViBa project from `C:\ArcGISByExample\telzaviba\Code` if you have closed it.

2. Double-click on `frmRealTime`, and then go to **View**, and click on **Code**.

3. Modify the `RunManeuver` method to the following; this is the piece of code that we wrote in *Chapter 3, Mapping Signal Strength*:

```
Public Sub RunManeuver(X As Double, Y As Double)
            Dim pPoint As IPoint = New Point
            pPoint.PutCoords(X, Y)

            Dim pDocument As IMxDocument = My.ArcMap.Application.
Document
            pDocument.ActiveView.GraphicsContainer.
DeleteAllElements()

            Dim pElement As IElement = New MarkerElement
            pElement.Geometry = pPoint
            Dim pMarkerElement As IMarkerElement = pElement

            Dim pPictureMarkerSymbol As IPictureMarkerSymbol = New
PictureMarkerSymbol
            pPictureMarkerSymbol.CreateMarkerSymbolFromFile(esriI
PictureType.esriIPicturePNG, "C:\ArcGISByExample\telzaviba\Icons\
Cell-phone.png")
            pPictureMarkerSymbol.Size = 20
            pMarkerElement.Symbol = pPictureMarkerSymbol

            pDocument.ActiveView.GraphicsContainer.
AddElement(pElement, 0)


            Dim pTowerLayer As IFeatureLayer = pDocument.FocusMap.
Layer(0)
            Dim pFeatureCursor As IFeatureCursor = pTowerLayer.
FeatureClass.Search(Nothing, False)
            Dim pFeature As IFeature = pFeatureCursor.NextFeature

            Dim pProximityOperator As IProximityOperator = pPoint

            Dim closestTowerOID As Long
            Dim shortestdistance As Double = Double.MaxValue

            Do Until pFeature Is Nothing

                Dim distance As Double = pProximityOperator.
ReturnDistance(pFeature.Shape)
                    If distance < shortestdistance Then
                        shortestdistance = distance
```

```
                    closestTowerOID = pFeature.OID
                End If

            pFeature = pFeatureCursor.NextFeature
        Loop

        Dim pClosestTower As IFeature = pTowerLayer.
FeatureClass.GetFeature(closestTowerOID)

        Dim pScreenDisplay As IScreenDisplay = pDocument.
ActiveView.ScreenDisplay
        pScreenDisplay.StartDrawing(pScreenDisplay.hDC, ESRI.
ArcGIS.Display.esriScreenCache.esriNoScreenCache)

        Dim pMarkerSymbol As IMarkerSymbol = New
SimpleMarkerSymbol
        Dim pColor As IRgbColor = New RgbColor
        pColor.RGB = RGB(255, 0, 0)
        pMarkerSymbol.Color = pColor
        pMarkerSymbol.Size = 10
        Dim pSymbol As ISymbol = pMarkerSymbol
        pSymbol.ROP2 = esriRasterOpCode.esriROPNotXOrPen
        pScreenDisplay.SetSymbol(pSymbol)
        pScreenDisplay.DrawPoint(pClosestTower.Shape)
        Threading.Thread.Sleep(200)
        pScreenDisplay.DrawPoint(pClosestTower.Shape)
        Threading.Thread.Sleep(200)
        pScreenDisplay.DrawPoint(pClosestTower.Shape)
        Threading.Thread.Sleep(200)
        pScreenDisplay.DrawPoint(pClosestTower.Shape)
        pScreenDisplay.FinishDrawing()
        Dim t As String = pClosestTower.Value(pClosestTower.
Fields.FindField("TOWER_ID"))
        Dim r As Double = pClosestTower.Value(pClosestTower.
Fields.FindField("TOWER_RANGE_METERS"))
        Dim s As Double = Math.Round((r - shortestdistance) /
r * 100)
        Dim pTextPoint As IPoint = pPoint
        pTextPoint.Y = pTextPoint.Y - 20
        Dim pTextElement As ITextElement = New TextElement
        pTextElement.Text = t & "(" & s & "%)"
        Dim pTextSymbol As ITextSymbol = New TextSymbol
        pTextSymbol.Size = 5
```

```
                    Dim pTheTextElement As IElement = pTextElement
                    pTheTextElement.Geometry = pTextPoint
                    pDocument.ActiveView.GraphicsContainer.
      AddElement(pTheTextElement, 0)

                    pDocument.ActiveView.Refresh()

            End Sub
```

4. Build your solution and run the `telzaviba_withbasemap_require_ internet.mxd` document.

5. Click on the **Simulate RealTime Cell Tower Maneuvering** button.

6. From the `frmRealTime` form, type 1 in the **Refresh Interval** textbox to speed up the refresh interval. Click on **Load GPS File** and then select the `C:\ArcGISByExample\telzaviba\Data\GPS_LOG_zigzag.txt` file.

7. Watch how the cell phone moves as it connects from one tower to another tower, while displaying the signal strength. Take a look at the following screenshot:

8.   Close ArcMap and choose not to save any changes.

You can further refactor the code by moving some segments to a method like we did with `RunManeuver`.

With that, we have completed our first example – the **TelZaViBa Cell Tower Analysis** tool. The tool is now very useful for engineers to help them find weak signal areas. You can find the latest code for this chapter under `B04847_04_code\telzaviba\FinalCode\TelZaviBa`.

 Exercise: Have the tool report the weakest signal and its location by adding a red mark on the map at the end of the simulation.

Summary

In this chapter, you managed to complete the first example, the TelZaViBa tower analysis tool. The tool which you just started working on in *Chapter 2, App 1 – the Cell Tower Analysis Tool*, and enhanced in *Chapter 3, Mapping Signal Strength*, has finally come into use in this chapter when you added the real-time maneuvering. You learned how to read an external text file with GPS values and convert these GPS values into projected map points. Then you used the code that you wrote in *Chapter 3, Mapping Signal Strength*, to draw the cell phone and execute the logic of finding the closest tower and signal strength. You also configured a timer interval to read the GPS records. Finally, you managed to highlight the closest tower as the simulation took place, which gives the user of the tool a clear idea of the signal strength and the connected cell tower at that particular location.

In the next chapter, you will start working on the restaurant management application using the extending ArcObjects approach, which is a bit different to add-ins but is still being used actively in the GIS community.

5
App 2 – Extending ArcObjects

In the last three chapters, you worked on a cell tower analysis tool that allowed us to introduce ArcGIS and ArcGIS add-ins. We scratched the surface of ArcObjects developments and we added few controls. This chapter, and the next two, will be dedicated to a more challenging example: the restaurants mapping application. We will approach this example with a slightly different approach in development using the extending ArcObjects method. In this chapter, we will work more closely with geodatabase structures.

In this chapter, we will discuss the following topics:

- Extending ArcObjects
- Adding the restaurants mapping toolbar
- Adding the restaurants viewer button
- Querying restaurants subtypes
- Finding restaurants in a subtype

Extending ArcObjects

The extending ArcObjects share the same underlying programming API with the add-in approach. However, they differ in the approach of deploying the application and at the level of flexibility offered to customize ArcGIS. With extending ArcObjects, you usually have to do more work. For example, setting up the toolbar and button in the add-in approach is done with a wizard. However, in the extending ArcObjects approach, you will need to write code to set up your toolbar. Despite that extra work, there is rewarding part in extending ArcObjects, and that is the degree of the flexibility opened to the developer to do great stuff, as we will see in *Chapter 7, Advanced Searching*.

Before we dive into the coding part, we need to do some preparation for our new project.

Preparing the geodatabase and map for bestaurants

Belize is thriving from tourism. Lots of tourists go there on holidays to enjoy its beautiful beaches and a wide range of restaurants. The government of Belize is trying to enrich tourists' experience in finding their favorite restaurants in the country more effectively.

To accomplish that, a new project titled bestaurants" has been proposed to design a restaurants mapping application on top of ArcGIS for Desktop to feature the best restaurants in Belize. The application will contain a map that shows the city of Belize and the restaurants with key icons based on the restaurant type. For example, a cafe will be shown as a coffee mug and a restaurant will be displayed as a knife and fork. Users should be able to search for restaurants by name, region, category, or rating.

The bestaurants team has provided us with the geodatabase and the map document, so we will simply copy the necessary files to your drive. Follow these steps to start your preparation of the data and map:

1. Copy the entire `bestaurants` folder in the supporting files for this chapter `B04847_05_Files\bestaurants\` to the `C:\ArcGISByExample\`.

2. Run the `bestaurants.mxd` file under the `C:\ArcGISByExample\ bestaurants\Data\bestaurants.mxd`. This should point to the geodatabase which is located under `C:\ArcGISByExample\bestaurants\Data\ bestaurants.gdb`, as illustrated in the following screenshot:

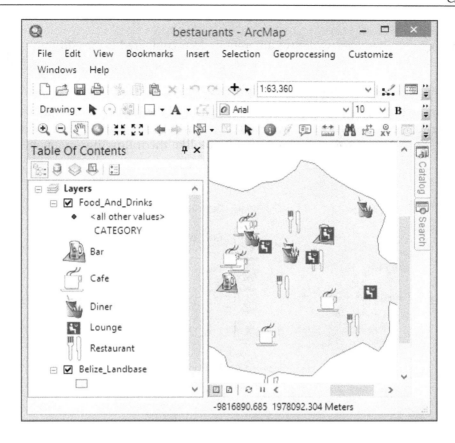

3. Close ArcMap and choose not to save any changes.

Preparing the bestaurants project

We will now start our project. First, we need to create our bestaurants Visual Studio by extending ArcObjects project. To do so, follow these steps:

1. From the Start menu, run Visual Studio Express 2013.

2. Go to the **File** menu | **New Project**.

3. Expand the **Templates** node, then **Visual Basic**, then **ArcGIS**, and then click on **Extending ArcObjects**. The list of projects will be displayed on the right.

Note the variety of different projects compared to the add-ins.

4. Select **Class Library (ArcMap)** project.

5. In the **Name** field, type `Bestaurants`, and in the location, browse to `C:\`
    `ArcGISByExample\bestaurants\Code`; if the `Code` folder is not there, create
    it. Your **New Project** dialog should look like the following screenshot:

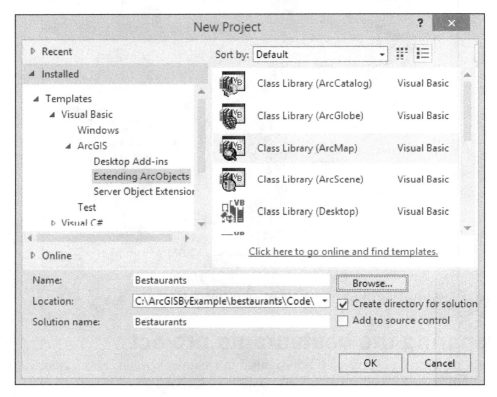

6. Click on **OK**.

7. In **ArcGIS Project Wizard**, you will be asked to select the references
    libraries you will need in your project. I always recommend selecting all the
    referencing, and then at the end of your project remove the unused ones.
    So, go ahead and right-click on **Desktop ArcMap** and click on **Select All**, as
    shown in the following screenshot:

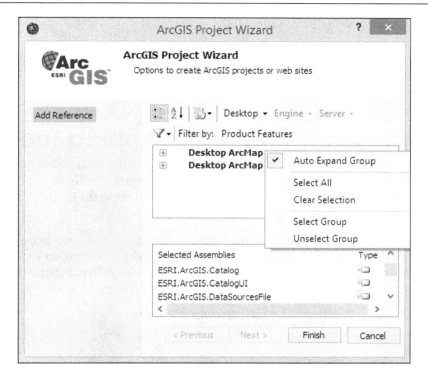

8. Click on **Finish** to create the project. It will take a while to add all references to your project.

9. Once your project is created, you will see one class called `Class1` has been added; we won't need it, so right-click on it and choose **Delete**. Then, click on **OK** to confirm, as shown in the following screenshot:

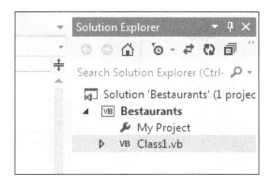

10. Go to **File** and click on **Save All**.

11. Exit the Visual Studio application.

You have finished preparing your Visual Studio with extending ArcObjects support. Move to the next section to write some code.

Adding the restaurants mapping toolbar

The thing I like about extending ArcObjects is the flexibility. You will notice how open this approach is when you start working with it more closely in the coming pages. The toolbar here is actually a class, which we have to add to the project. To add the toolbar, follow these steps:

1. Open Visual Studio Express in administrator mode; we need to do this since our project is actually writing to the registry this time, so it needs administrator permissions. To do that, right-click on **Visual Studio** and click on **Run as administrator**, as shown in the following screenshot:

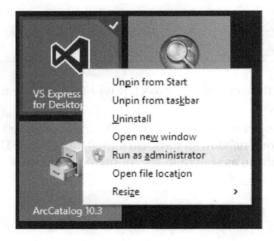

2. Go to **File**, then click on **Open Project**, browse to the Bestaurants project from the C:\ArcGISByExample\bestaurants\Code, and click on **Open**.

3. Click on the Bestaurants project from **Solution Explorer** to activate it.

4. From the **Project** menu, click on **Add Class**.

5. Expand the **ArcGIS** node and then click on the **Extending ArcObjects** node.

6. Select **Base Toolbar** and name it tbBestaurants.vb, as illustrated in the following screenshot:

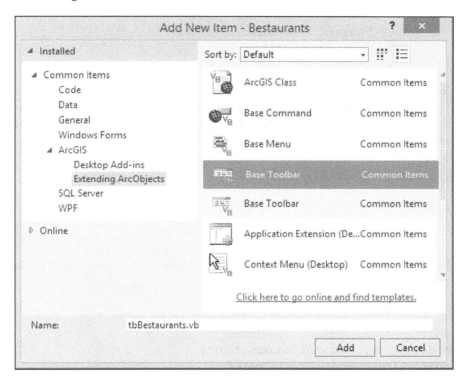

7. Click on **Add** to open **ArcGIS New Item Wizard Options**.

8. From **ArcGIS New Item Wizard Options**, select **Desktop ArcMap** since we will be programming against ArcMap. Click on **OK**.

9. The tbBestauraunts.vb class has been added to your project. We need to note a few important key elements. Take a look at the ProgID property Bestaurants.tbBestaurants. This is the unique identifier of your class, which is composited of the projectname.classname. We will be using ProgID a lot in this type of programming.

10. The `Caption` property is what is displayed when the toolbar loads. It currently defaults to **MY VB.Net Toolbar**; change it to **Bestaurants Toolbar** as follows:

```
Public Overrides ReadOnly Property Caption() As String
        Get
            'TODO: Replace bar caption
            Return "Bestaurants Toolbar"
        End Get
    End Property
```

11. Your toolbar is currently empty, which means it doesn't have buttons or tools. We can actually add built-in out-of-the-box buttons by using their corresponding ProgID and reuse their functionality. Go to the `New` method and replace all code there with the following code. This will add the zoom in tool:

```
Public Sub New()

AddItem("esriArcMapUI.ZoomInTool")

End Sub
```

12. We just added the zoom out, pan, and full extent tool. Let us add a couple more commands to our toolbar as follows:

```
Public Sub New()

AddItem("esriArcMapUI.ZoomInTool")
AddItem("esriArcMapUI.ZoomOutTool")
AddItem("esriArcMapUI.PanTool")
AddItem("esriArcMapUI.FullExtentCommand")

End Sub
```

13. Now it is time to test our new toolbar. Point to **Build** and then click on **Build Solution**. Make sure ArcMap is not running. If you got an error, make sure you have run the Visual Studio as administrator.

> For a list of all ArcMap commands, refer to http://bit.ly/b04748_arcmapids. Check the commands with namespace esriArcMapUI.

14. Run `bestaurants.mxd` in `C:\ArcGISByExample\bestaurants\Data\`
    `bestaurants.mxd`.

15. From ArcMap, point to **Customize** menu, then **Toolbars**, and then select the
    Bestaurants Toolbar we just created. You should see the toolbar popup on
    ArcMap with the four added commands, as shown in the following screenshot:

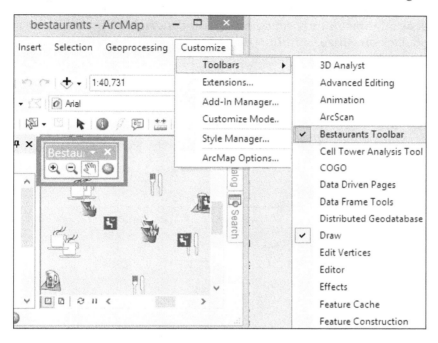

16. Close ArcMap and choose not to save any changes.

Now, that we have shown the toolbar and added some built-in commands on it.
Now we will create and add our own button to the toolbar.

Adding the restaurants viewer button

In this section, we will add an ArcMap button to the project. We will add logic to
this button so that if the user clicks on it, it will display a form. This form will be the
restaurants viewer form where all restaurants searches are conducted.

Adding the button

We will now add a button to our project; a button in extending ArcObjects is known as a command. To add the command, follow these steps:

1. If necessary, open Visual Studio Express in administrator mode; we need to do this since our project is actually writing to the registry this time, so it needs administrator permissions. To do that, right-click on **Visual Studio** and click on **Run as administrator**.

2. Go to **File**, then click on **Open Project**, browse to the Bestaurants project from the C:\ArcGISByExample\bestaurants\Code, and click on **Open**.

3. Click on the Bestaurants project from the Solution Explorer to activate it.

4. From the **Project** menu, click on **Add Class**.

5. Expand the **ArcGIS** node and then click on the **Extending ArcObjects** node.

6. Select **Base Command** and name it cmViewer.vb, as illustrated in the following screenshot:

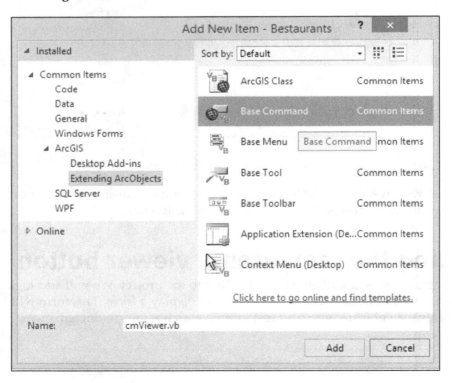

7. Click on **Add** to open **ArcGIS New Item Wizard Options**.

8. From **ArcGIS New Item Wizard Options**, select **Desktop ArcMap Command** since we will be programming against ArcMap. Click on **OK**.

9. Take note of the `Bestaurants.cmViewer` ProgID. We will use this ProgID in the next section to add the button to the toolbar just as we added the built-in out-of-the-box buttons.

10. If necessary, double-click on the `cmViewer.vb` to edit it.

11. In the `New` method, update the properties of the command as follows. This will update the name and caption and other properties of the command. There is a piece of code that loads the command icon. Leave that as it is:

```
Public Sub New()
MyBase.New()

        ' TODO: Define values for the public properties
MyBase.m_category = "Bestaurants"  'localizable text
MyBase.m_caption = "Bestaurants Viewer"    'localizable text
MyBase.m_message = "Bestaurants Viewer"    'localizable text
MyBase.m_toolTip = "Bestaurants Viewer" 'localizable text
MyBase.m_name = "Bestaurants_BestaurantsViewer"

        Try
            'TODO: change bitmap name if necessary
            Dim bitmapResourceName As String = Me.GetType().Name +
".bmp"
MyBase.m_bitmap = New Bitmap(Me.GetType(), bitmapResourceName)
        Catch ex As Exception
System.Diagnostics.Trace.WriteLine(ex.Message, "Invalid Bitmap")
        End Try

    End Sub
```

12. Now we need to add the Bestaurants viewer from which we will have the restaurants category drop-down list for starters. Point to **Project** and then click on **Add Windows Form**.

13. Name the form `frmRestaurantsViewer.vb` and click on **Add**.

14. Use **Form Designer** to add and set the following controls. Two labels, name them lblCategory and lblVenue, and two **ComboBoxes**, name then cmbCategory and cmbVenue, respectively, as shown in the following screenshot. You can set the **Name** property by clicking on the control and editing the **Name** property:

15. Double-click on the cmViewer.vb to edit it.

16. Scroll to the OnClick method. This method will be executed when the user clicks on the command. Let us add a simple code to show our new form:

```
Public Overrides Sub OnClick()
    'TODO: Add cmViewer.OnClick implementation

    Dim frmRestaurantsViewer As New frmRestaurantviewer
frmRestaurantsViewer.Show()

    End Sub
```

17. One last change before we build our solution: we need to change the default icon of our command. To do that, double-click on cmViewer.bmp to open the picture editor, replace the picture with the bestaurants.bmp, which can be found under C:\ArcGISByExample\bestaurants\icons\bestaurants. bmp, and save as cmViewer.bmp.

18. Save your project and move to the next step to assign the command to the toolbar.

Assigning a button to the toolbar

Adding our new button to the toolbar is easy. All that remains is taking the ProgID of our command and adding it to the toolbar. Follow these steps to do so:

1. If necessary, open Visual Studio Express in administrator mode; we need to do this since our project is actually writing to the registry this time, so it needs administrator permissions. To do that, right-click on **Visual Studio** and click on **Run as administrator**.

2. Go to **File**, then click on **Open Project**, browse to the Bestaurants project from the **C:\ArcGISByExample\bestaurants\Code**, and click on **Open**.

3. Double-click on tbBestaurants.vb to edit it.

4. On the New method, remove all the AddItem calls and add our command's ProgID instead, as explained in the following code:

```
Public Sub New()

AddItem("Bestaurants.cmViewer")

End Sub
```

5. Now it is time to test our new command. Go to **Build** and then click on **Build Solution**. Make sure ArcMap is not running. If you got an error, make sure you have run Visual Studio as administrator.

6. Run bestaurants.mxd in C:\ArcGISByExample\bestaurants\Data\ bestaurants.mxd.

7. Note that now we have only one command on our bestaurants toolbar and it has a restaurants icon.

8. Click on the **Restaurants viewer** command to show the restaurants viewer form, as illustrated in the following screenshot:

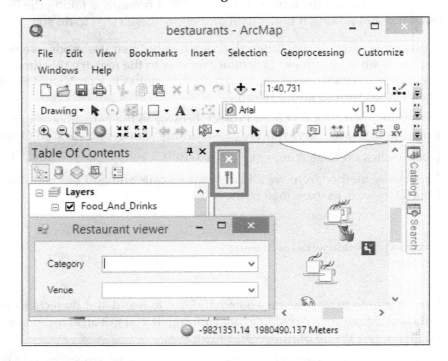

9. Close ArcMap and choose not to save any changes.

Up until this stage, we were only preparing the project. We didn't actually write any ArcObjects code. Move to the next section to learn how to populate the drop-down list with some ArcObjects.

Querying the subtypes of restaurants

To learn how to query the geodatabase in general, we need to learn about the workspaces that are the geodatabase connections. The workspace is used to establish a channel between the client and the geodatabase. Once this channel is established, the client can query, write, edit, and delete features in the geodatabase.

A geodatabase workspace

A workspace is a unique geodatabase connection established by the user in order to query or edit the geodatabase. A workspace can also refer to other data structures such as shape files, coverages, CAD files, and more. We will start this section by learning how to open a workspace from within the code. To create a workspace in ArcObjects, we require a **workspace factory** object. The workspace factory determines the type of workspace you want to create. For instance, in our case, we have a file geodatabase, so we will need a file geodatabase workspace factory.

Workspace is the connection to the geodatabase; ArcObjects uses this to query and update the geodatabase.

Workspace factory is the object used to create and establish geodatabase connections; there are many types of factories depending on the type of the geodatabase.

For more information about workspace and workspace factories, refer to `http://bit.ly/b04748_workspace`.

In this exercise, we will open a workspace to our `bestaurants.gdb` file, and then open a feature class that has all these restaurants and simply display the number of restaurants in that feature class. Follow these steps:

1. If necessary, open Visual Studio Express in administrator mode; we need to do this since our project is actually writing to the registry this time, so it needs administrator permissions. To do that, right-click on **Visual Studio** and click on **Run as administrator**.

2. Go to **File**, then click on **Open Project**, browse to the `Bestaurants` project from the `C:\ArcGISByExample\bestaurants\Code`, and click on **Open**.

3. Right-click on `frmRestaurantsViewer.vb` and click on **View Code** to view its code.

4. Add the following code to create the **Form Load** event, which is the code that will be executed when the form loads:

```
Public Class frmRestaurantviewer

    Private Sub frmRestaurantviewer_Load(sender As Object, e As
EventArgs) Handles MyBase.Load

    End Sub

End Class
```

5. To use the workspace, we will need to import some libraries. Add the following import statements at the top of your class:

```
Imports ESRI.ArcGIS.DataSourcesGDB
Imports ESRI.ArcGIS.Geodatabase
```

6. We will now open a workspace connection to `bestaurants.gdb` located on `C:\ArcGISbyExample\bestaurants\data\bestaurants.gdb`. Write the following code to open the workspace in the `frmRestaurantviewer_Load` method, which will be executed when the form first loads. We will use the `OpenFromFile` method that takes the path of the geodatabase. Note that we created `FileGDBWorkspaceFactory` since we are dealing with a file geodatabase:

```
Private Sub frmRestaurantviewer_Load(sender As Object, e As
EventArgs) Handles MyBase.Load
        Dim pWorkspaceFactory As IWorkspaceFactory = New
FileGDBWorkspaceFactory
        Dim pWorkspace As IWorkspace = pWorkspaceFactory.
OpenFromFile("c:\ArcGISbyExample\bestaurants\data\bestaurants.
gdb", Me.Handle.ToInt32)

    End Sub
```

7. Now that we have the workspace, we can open the feature class `Food_and_Drinks`, which has all the restaurants. Write the following code to open the feature class:

```
Private Sub frmRestaurantviewer_Load(sender As Object, e As
EventArgs) Handles MyBase.Load
        Dim pWorkspaceFactory As IWorkspaceFactory = New
FileGDBWorkspaceFactory
        Dim pWorkspace As IWorkspace = pWorkspaceFactory.
OpenFromFile("c:\ArcGISbyExample\bestaurants\data\bestaurants.
gdb", Me.Handle.ToInt32)

        Dim pFeatureWorkspace As IFeatureWorkspace = pWorkspace
        Dim pFeatureClass As IFeatureClass = pFeatureWorkspace.
OpenFeatureClass("Food_and_Drinks")

    End Sub
```

8. For testing only, let us display a message with the total number of features inside that feature class. Write the following code to do so:

```
Private Sub frmRestaurantviewer_Load(sender As Object, e As
EventArgs) Handles MyBase.Load
        Dim pWorkspaceFactory As IWorkspaceFactory = New
FileGDBWorkspaceFactory
```

```
        Dim pWorkspace As IWorkspace = pWorkspaceFactory.
OpenFromFile("c:\ArcGISbyExample\bestaurants\data\bestaurants.
gdb", Me.Handle.ToInt32)

        Dim pFeatureWorkspace As IFeatureWorkspace = pWorkspace
        Dim pFeatureClass As IFeatureClass = pFeatureWorkspace.
OpenFeatureClass("Food_and_Drinks")

MsgBox(pFeatureClass.FeatureCount(Nothing))
        End Sub
```

9. Build and run the `bestaurants.mxd` file.

10. Click on the **Restaurants viewer** button to show the restaurants viewer; this will display a message of the total number of restaurants. We can confirm that this number is true by right-clicking on the `Food_and_Drinks` layer and clicking on **Open attribute table**, as illustrated in the following screenshot:

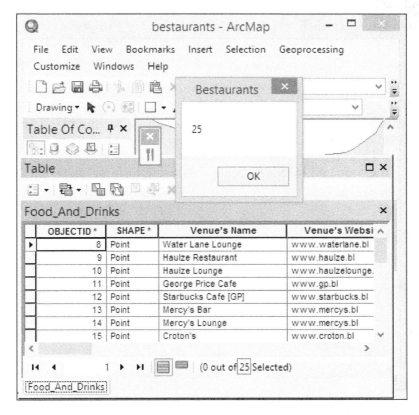

11. Close ArcMap and choose not to save the changes.

Populating subtypes

In this section, we will work with subtypes. We will use the subtypes that are defined in the feature class and populate them in a drop-down list.

 Subtype: A subset of features in a given feature class sharing the same attributes.

If you have noticed in the `Food_and_Drinks` layer, there are many categories in the restaurants. In this section, we will query those categories and populate them in the categories drop-down list in the form. Follow these steps:

1. If necessary, open Visual Studio Express in administrator mode; we need to do this since our project is actually writing to the registry this time, so it needs administrator permissions. To do that, right-click on **Visual Studio** and click on **Run as administrator**.

2. Go to **File**, then click on **Open Project**, browse to the `Bestaurants` project from the `C:\ArcGISByExample\bestaurants\Code`, and click on **Open**.

3. Right-click on `frmRestaurantsViewer.vb` and click on **View Code** to view its code.

4. Remove the message box that displays the number of features.

5. Since this is a coded value subtype, we will be expecting two values with each subtype: the code and the description. Now we will loop through the subtypes and then populate them in a key value pair dictionary with the key being the code and the value being the description as follows:

```
Dim pFeatureClass As IFeatureClass = pFeatureWorkspace.
OpenFeatureClass("Food_and_Drinks")

Dim pSubtypes As ISubtypes = pFeatureClass
        Dim eSubtypes As IEnumSubtype = pSubtypes.Subtypes

eSubtypes.Reset()
        Dim rCode As Integer
        Dim rName As String
cmbCategory.Items.Clear()

        Dim dicCategories As New Dictionary(Of Integer, String)

rName = eSubtypes.Next(rCode)
        Do While rName<> ""
dicCategories.Add(rCode, rName)
```

```
rName = eSubtypes.Next(rCode)
        Loop


cmbCategory.DisplayMember = "Value"
cmbCategory.ValueMember = "Key"
cmbCategory.DataSource = New BindingSource(dicCategories, Nothing)

    End Sub
```

6. Build and run `bestaurants.mxd`.

7. Click on the **Restaurants viewer** command to show the restaurants; you will notice that the **Category** drop-down list has been populated with the subtypes, as illustrated in the following screenshot:

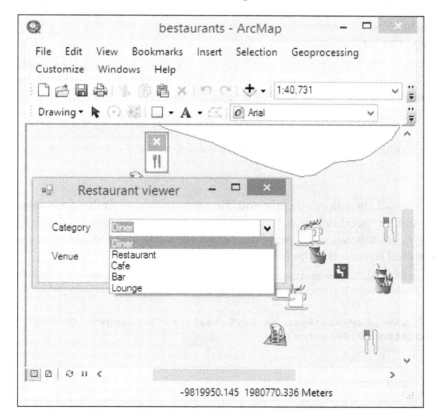

8. Close ArcMap and choose not to save any changes.

Finding restaurants in a subtype

We have populated the subtypes of the restaurants in the **Category** drop-down list. In this section, we will learn how to query the feature class and filter, so we only display restaurants of particular category. For instance, if the user selects **Cafe** from the **Category** list, only cafes will be populated in the venues list.

Follow these steps to add the query:

1. If necessary, open Visual Studio Express in administrator mode; we need to do this since our project is actually writing to the registry this time, so it needs administrator permissions. To do that, right-click on **Visual Studio** and click on **Run as administrator**.

2. Go to **File**, then click on **Open Project**, browse to the Restaurants project from the `C:\ArcGISByExample\bestaurants\Code`, and click on **Open**.

3. Right-click on `frmRestaurantsViewer.vb` and click on **View Designer** to view the form designer.

4. We want to write logic to populate the venues when the user selects a new category that is on `cmbCategory_SelectedIndexChanged`. You can double-click on `cmbCategory` to add the method or you can simply write it as follows:

    ```
        Private Sub cmbCategory_SelectedIndexChanged(sender As Object, e
    As EventArgs) Handles cmbCategory.SelectedIndexChanged


        End Sub
    ```

5. We saved the category code and the category name in the drop-down list, and in the geodatabases, the code is saved, which means we have to query the geodatabases with the code. To get the code and the name from the selected item, write the following code:

    ```
        Private Sub cmbCategory_SelectedIndexChanged(sender As Object, e
    As EventArgs) Handles cmbCategory.SelectedIndexChanged


        Dim pKeyValuePair As KeyValuePair(Of Integer, String) =
    cmbCategory.SelectedItem


        End Sub
    ```

6. Next we add the following code to open the feature class:

```
 Private Sub cmbCategory_SelectedIndexChanged(sender As Object, e
As EventArgs) Handles cmbCategory.SelectedIndexChanged
        Dim pKeyValuePair As KeyValuePair(Of Integer, String) =
cmbCategory.SelectedItem

        Dim pWorkspaceFactory As IWorkspaceFactory = New
FileGDBWorkspaceFactory
        Dim pWorkspace As IWorkspace = pWorkspaceFactory.
OpenFromFile("c:\ArcGISbyExample\bestaurants\data\bestaurants.
gdb", Me.Handle.ToInt32)

        Dim pFeatureWorkspace As IFeatureWorkspace = pWorkspace
        Dim pFeatureClass As IFeatureClass = pFeatureWorkspace.
OpenFeatureClass("Food_and_Drinks")

    End Sub
```

7. Now we need to use the search function on the feature class to query and pass a query filter. The filter is Category = CategoryCode, as shown:

```
 Private Sub cmbCategory_SelectedIndexChanged(sender As Object,
e As EventArgs) Handles cmbCategory.SelectedIndexChanged
        Dim pKeyValuePair As KeyValuePair(Of Integer, String) =
cmbCategory.SelectedItem

        Dim pWorkspaceFactory As IWorkspaceFactory = New
FileGDBWorkspaceFactory
        Dim pWorkspace As IWorkspace = pWorkspaceFactory.
OpenFromFile("c:\ArcGISbyExample\bestaurants\data\bestaurants.
gdb", Me.Handle.ToInt32)

        Dim pFeatureWorkspace As IFeatureWorkspace = pWorkspace
        Dim pFeatureClass As IFeatureClass = pFeatureWorkspace.
OpenFeatureClass("Food_and_Drinks")

        Dim pQFilter As IQueryFilter = New QueryFilter
pQFilter.WhereClause = "CATEGORY = " &pKeyValuePair.Key

        Dim pFeatureCursor As IFeatureCursor = pFeatureClass.
Search(pQFilter, False)

    End Sub
```

8. Finally, we populate the venues drop-down list with features using the object ID as the key, and the name of the restaurant as the value:

```
Dim pFeature As IFeature = pFeatureCursor.NextFeature
     Dim dictVenues As New Dictionary(Of Integer, String)


     Do Until pFeature Is Nothing

dictVenues.Add(pFeature.OID, pFeature.Value(pFeature.Fields.
FindField("NAME")))
pFeature = pFeatureCursor.NextFeature
     Loop


cmbVenue.DisplayMember = "Value"
cmbVenue.ValueMember = "Key"
cmbVenue.DataSource = New BindingSource(dictVenues, Nothing)

End Sub
```

9. Build and run `bestaurants.mxd`.

10. Click on the **Restaurants viewer** command to show the restaurants and select a **Category**, say **Cafe**; note that the venues will be populated with only cafes, as illustrated in the following screenshot:

11. Close ArcMap and choose not to save any changes.

This is the end of the chapter. You can find the latest code under `B04847_05_Files\`
`bestaurants\FinalCode\`.

Summary

In this chapter, you started writing the restaurants mapping application, which is called Bestaurants. In the first half of the chapter, you spent time learning to use the extending ArcObjects approach which is a bit different from the add-in to create the project that had Bestaurants Toolbar and then added command to it. Then, in the second half of the chapter, you learned about workspaces and workspace factories that helped establishing connection to the geodatabase in order to query it. You then used these skills to query the restaurants and populate them on the form.

In the next chapter, you will keep enhancing the restaurants mapping application to add more functionality to it. You will learn about geodatabase relationships and use them to query related data.

6
Reviews and Ratings

In the last chapter, we initiated the Bestaurants project. We learned about the extending ArcObjects approach and how we can use it as a new approach in development besides add-ins. We used ArcObjects to work with workspaces and do simple queries in order to query restaurants and populate them in the form. This chapter will introduce us to the relationship queries in geodatabases, and interacting with the map.

In this chapter, we will discuss the following topics:

- Introducing relationships
- Querying reviews and ratings
- Highlighting restaurants
- Filtering restaurants on the map

Introducing relationships

Relationships are the core of any relational database management system. The relational database model is based on multiple tables connected via relation. Relationships come in many forms, one to one, one to many, and many to many. In our case, a certain venue has many reviews which make this relationship a one-to-many relationship.

The reviews and ratings table

There can be many reviews and ratings for a given restaurant, and putting this information in the feature class will create duplicate information. This is why a related table can be useful here. Both the reviews and ratings are stored in a related table called VENUES_REVIEW. Follow these steps to inspect this related table:

1. If necessary, copy the entire bestaurants folder from the supporting files for this chapter, B04847_06_Files\bestaurants\ to the C:\ArcGISByExample\. If you already have done this in *Chapter 5, App 2 – Extending ArcObjects*, you can skip this step.

2. From the Start menu, run ArcCatalog.

3. Connect to the C:\ArcGISByExample\ folder and browse to C:\ArcGISbyExample\bestaurants\Data\Bestaurants.gdb, as shown in the following screenshot:

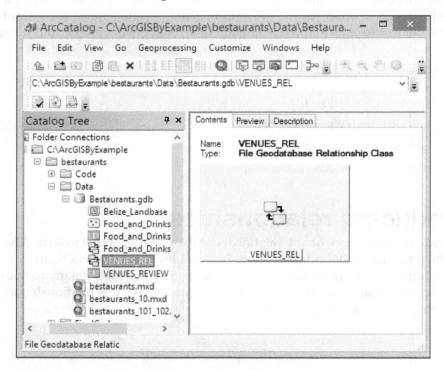

4. Right-click on the VENUES_REL relationship class and then click on **Properties**.

5. Take note of the **Cardinality** that is **1 – M**, which indicates a one-to-many relationship. Note that the origin feature class is **Food_and_Drinks** and the destination table is **VENUES_REVIEWS**, which has all the reviews, as illustrated in the following screenshot:

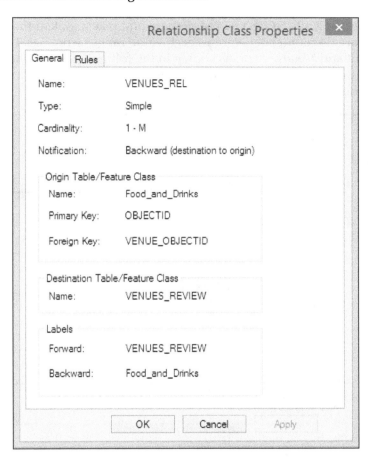

6. Click on **OK** to close the **Relationship Class Properties** dialog.

7. Click on the **VENUES_REVIEW** table, and then from the right-hand side, click on **Preview** to view the underlying records.

8. Take note of the **VENUE_OBJECTID** field, which is the object ID of the restaurant. Note how some restaurant IDs have been duplicated here, which indicates that there are multiple reviews for the same restaurant.

9. Scroll horizontally to the **RATING** field. Note that there are multiple values: **Excellent**, **Good**, **Average**, **Fair**, and **Poor**.

10. Right-click on the Bestaurants.gdb database and click on **Properties**.

11. From the **Database Properties** dialog, click on **Domains**.

12. Click on the **RATING_SYSTEM** domain and take a look at the values. These values are domain-coded values, each value ranges from 5 to 1, with 5 being excellent and 1 being poor, as shown in the following screenshot. We will use these values later to calculate the average rating for the restaurant by adding up all the ratings and dividing it by the number of reviews:

13. Click on **OK** to close the **Database Properties** dialog.

14. Close ArcCatalog.

Querying reviews and ratings

Here we will learn how to query related records and use this knowledge to do some calculations on these records.

Retrieving related records

In this section, we will learn how to query the related reviews table and show them on our form. For that, we need to first select a particular restaurant from the list and then query the table:

1. Open Visual Studio Express in Administrator mode; we need to do this since our project is actually writing to the registry this time, so it needs administrator permissions. To do that, right-click on Visual Studio and click on **Run as administrator**.

2. Go to **File | Open Project**, browse to the `Bestaurants` project from the `C:\ArcGISByExample\bestaurants\Code`, and click on **Open**.

> You can start from where you left in *Chapter 5, App 2 – Extending ArcObjects*, or you can use the code `B04847_06_Files\bestaurants\StartCode`, copy it in to `C:\ArcGISByExample\bestaurants\Code`, and start working.

3. Double-click on `frmRestaurantsViewer.vb` to open the form designer.

4. From the Toolbox, add two labels (`lblRatingHeader` and `lblAvgRating`) to display the rating of the selected venue. Set the **Text** property of `lblRatingHeader` to **Average Rating** and set the **Text** property of `lblAvgRating` to **X**. The `lblAvgRating` label will be populated from the code. Drag the **ListBox** control and add it to the form. The list will be used to display the reviews. Name the list `lstReviews`, as shown in the following screenshot:

5. We want to list all the reviews for the selected venue when the user selects an item from the venue drop-down list. This means we have to add our code in to the **Venue**. Double-click on the **Venue** drop-down list to add the `cmbVenue_SelectedIndexChange` method, as shown in the following code:

```
Private Sub cmbVenue_SelectedIndexChanged(sender As Object, e As
EventArgs) Handles cmbVenue.SelectedIndexChanged

    End Sub
```

We will now need access to the workspace. We know how to open it but it is not a good practice to copy this code over. This is why we will create a new method called `GetVenuesFeatureClass`, which will take care of all this and return a feature class, as shown:

```
Private Function GetVenuesFeatureClass() As IFeatureClass
    Dim pWorkspaceFactory As IWorkspaceFactory = New
FileGDBWorkspaceFactory
    Dim pWorkspace As IWorkspace = pWorkspaceFactory.
OpenFromFile("c:\ArcGISbyExample\bestaurants\data\bestaurants.
gdb", Me.Handle.ToInt32)

    Dim pFeatureWorkspace As IFeatureWorkspace = pWorkspace
    Dim pFeatureClass As IFeatureClass = pFeatureWorkspace.
OpenFeatureClass("Food_and_Drinks")
    Return pFeatureClass
End Function
```

6. You can use this function whenever you require access to this feature class. Optionally, update your code to use this new function:

```
Private Sub frmRestaurantviewer_Load(sender As Object, e As
EventArgs) Handles MyBase.Load
    Dim pFeatureClass As IFeatureClass =
GetVenuesFeatureClass()

    Dim pSubtypes As ISubtypes = pFeatureClass
    Dim eSubtypes As IEnumSubtype = pSubtypes.Subtypes
    . .



    Private Sub cmbCategory_SelectedIndexChanged(sender As Object,
e As EventArgs) Handles cmbCategory.SelectedIndexChanged
    Dim pKeyValuePair As KeyValuePair(Of Integer, String) =
cmbCategory.SelectedItem
```

```
        Dim pFeatureClass As IFeatureClass =
GetVenuesFeatureClass()

        Dim pQFilter As IQueryFilter = New QueryFilter
pQFilter.WhereClause = "CATEGORY = " &pKeyValuePair.Key
```

7. Now we require access to the **VENUES_REVIEW** table, to write the `GetReviewsTable` function to return that table. We will use the `OpenTable` method in the workspace instead of `OpenFeatureClass`:

```
    Private Function GetReviewsTable() As ITable
        Dim pWorkspaceFactory As IWorkspaceFactory = New
FileGDBWorkspaceFactory
        Dim pWorkspace As IWorkspace = pWorkspaceFactory.
OpenFromFile("c:\ArcGISbyExample\bestaurants\data\bestaurants.
gdb", Me.Handle.ToInt32)

        Dim pFeatureWorkspace As IFeatureWorkspace = pWorkspace
        Dim pTable As ITable = pFeatureWorkspace.
OpenTable("VENUES_REVIEW")
        Return pTable
    End Function
```

8. The table can be treated similarly to a feature class, except it is a general case and doesn't have a shape geometry field. To query the table for reviews, we need to use a query filter and show only reviews for the selected restaurant. The select restaurant object ID is stored in the key pair, if you remember in *Chapter 5, App 2 – Extending ArcObjects*. Write the following code to execute the query in `cmbVenue_SelectedIndexChange`:

```
    Private Sub cmbVenue_SelectedIndexChanged(sender As Object, e
As EventArgs) Handles cmbVenue.SelectedIndexChanged
        Dim pSelectedVenueKeyValuePair As KeyValuePair(Of Integer,
String) = cmbVenue.SelectedItem

lstReviews.Items.Clear()

        Dim pTable As ITable = GetReviewsTable()
        Dim pQFilter As IQueryFilter = New QueryFilter
pQFilter.WhereClause = "VENUE_OBJECTID = "
&pSelectedVenueKeyValuePair.Key
        Dim pRowCursor As ICursor = pTable.Search(pQFilter, False)

    End Sub
```

Note that we have used ICursor instead of IFeatureCursor to retrieve the query for the table.

9. Now we will loop through the cursor and populate the lstReviews, with both the review, which can be found in the **REVIEW** column, and the rating, which is in the **RATING** column:

```
    Private Sub cmbVenue_SelectedIndexChanged(sender As Object, e
As EventArgs) Handles cmbVenue.SelectedIndexChanged
        Dim pSelectedVenueKeyValuePair As KeyValuePair(Of Integer,
String) = cmbVenue.SelectedItem

lstReviews.Items.Clear()

        Dim pTable As ITable = GetReviewsTable()
        Dim pQFilter As IQueryFilter = New QueryFilter
pQFilter.WhereClause = "VENUE_OBJECTID = "
&pSelectedVenueKeyValuePair.Key
        Dim pRowCursor As ICursor = pTable.Search(pQFilter, False)


        Dim pRow As IRow = pRowCursor.NextRow
        Do UntilpRow Is Nothing
            Dim rating As Integer = pRow.Value(pRow.Fields.
FindField("RATING"))
            Dim sreview As String = pRow.Value(pRow.Fields.
FindField("REVIEW"))

lstReviews.Items.Add(sreview& " [" & rating & "]")
pRow = pRowCursor.NextRow
        Loop
    End Sub
```

 Note that we have used IRow instead of IFeature to loop through the records since we are dealing with a table.

10. Go to **Build** | **Build Solution**; make sure ArcMap is not running. If you got an error, make sure you are running the Visual Studio as administrator.

11. Run `bestaurants.mxd` under
    `C:\ArcGISByExample\bestaurants\Data\bestaurants.mxd`.

12. Click on the **Bestaurants Viewer** button to view the **Bestaurants** form.

13. Select **Lounge** from **Category** and then select the **Water Lane Lounge** venue. You will see that there are two reviews: one saying **Amazing place** with a rating of 3 (average) and another one saying **Great** with a rating of 4, as shown in the following screenshot:

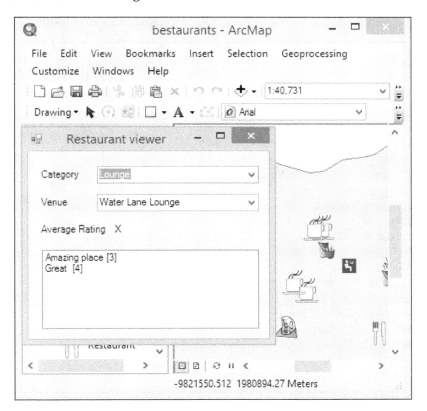

Calculating the average rating

The average rating of a given restaurant is the sum of all the review ratings divided by the number of reviews. We know that each review has one rating only, so this can be calculated easily. Follow these steps to edit your code for the rating calculation:

1. If necessary, open Visual Studio Express in Administrator mode; we need to do this since our project is actually writing to the registry this time, so it needs administrator permissions. To do that, right-click on Visual Studio and click on **Run as administrator**.

2. Go to **File | Open Project**, browse to the `Bestaurants` project from the `C:\ArcGISByExample\bestaurants\Code`, and click on **Open**.

3. Go to the `cmbVenue_SelectedIndexChanged` method and add the following code to the loop. `lReviewsCount` will count the total number of reviews, `lSumRating` will sum all ratings, and `dAvgRating` will have the average rating. We will also round it to the nearest decimal point:

```
Dim dAvgRating As Double = 0
Dim lSumRating As Integer = 0
Dim lReviewsCount As Integer = 0

Dim pRow As IRow = pRowCursor.NextRow
Do UntilpRow Is Nothing
    Dim rating As Integer = pRow.Value(pRow.Fields.
FindField("RATING"))
    Dim sreview As String = pRow.Value(pRow.Fields.
FindField("REVIEW"))

lstReviews.Items.Add(sreview& " [" & rating & "]")
lSumRating = lSumRating + rating
lReviewsCount = lReviewsCount + 1
pRow = pRowCursor.NextRow
Loop

dAvgRating = lSumRating / lReviewsCount
lblAvgRating.Text = Math.Round(dAvgRating, 1)
```

4. Go to **Build | Build Solution**; make sure ArcMap is not running. If you got an error, make sure you are running the Visual Studio as administrator.

5. Run `bestaurants.mxd` under
   `C:\ArcGISByExample\bestaurants\Data\bestaurants.mxd`.

6. Click on the **Bestaurants Viewer** button to view the **Bestaurants** form.

7. Select **Lounge** from **Category** and then select the **Water Lane Lounge** venue. You will see the reviews and also the average rating of 3.5, as shown in the following screenshot:

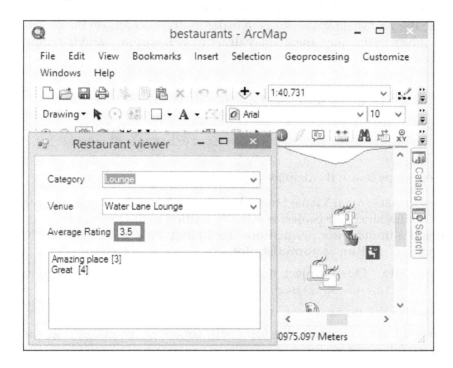

8. Close ArcMap and choose not to save any changes.

 This code will generate an error in one case. What is it? Can you fix it?

Highlighting restaurants

In this section, we will add a new button on our Bestaurants form to highlight a selected venue. To do that, we need to introduce the ArcGIS Display object that allows us to work with the screen display and manipulate it.

The ArcGIS Display object

We can use the ArcGIS Display object in order to draw objects on the screen. Unlike adding graphics to the map, the display drawing is temporary and is removed the moment you refresh.

 The ArcGIS Display object is an ArcObject class that allows interaction with the screen. It allows you to draw temporary symbols on the screen to simulate flashing or highlighting. To learn more about display object, refer to http://bit.ly/b04748_flashgeo.

Follow these steps to use the display object:

1. If necessary, open Visual Studio Express in Administrator mode; we need to do this since our project is actually writing to the registry this time, so it needs administrator permissions. To do that, right-click on Visual Studio and click on **Run as administrator**.

2. Go to **File | Open Project**, browse to the Bestaurants project from the C:\ArcGISByExample\bestaurants\Code, and click on **Open**.

3. Double-click on frmRestaurantsViewer to open the form designer.

4. Drag a Button control to the form and name it btnHighlight, as shown in the following screenshot:

5. Double-click on the `btnHighlight` button to open its underlying code as follows:

    ```
    Private Sub btnHighlight_Click(sender As Object, e As
    EventArgs) Handles btnHighlight.Click


    End Sub
    ```

6. To get access to the ArcGIS screen display, we need to get access to the ArcMap application object. Unlike ArcGIS add-ins, the application object cannot be accessed using the `My.ArcMap` keyword, instead it has to get passed in every class. Add the **Application** property to `frmRestaurantsviewer`, as shown:

    ```
    Private _application As ESRI.ArcGIS.Framework.IApplication
    Public Property ArcMapApp() As ESRI.ArcGIS.Framework.
    IApplication
        Get
            Return _application
        End Get
    Set(ByVal value As ESRI.ArcGIS.Framework.IApplication)
            _application = value
        End Set
    End Property
    ```

7. Double-click on our command `cmViewer.vb` to open the code.

8. Note private `m_application` as `IApplication`, which is the object we need to send to our form to use it. In the `OnClick` method, add the following code to pass the application object:

    ```
    Public Overrides Sub OnClick()
        'TODO: Add cmViewer.OnClick implementation


        Dim frmRestaurantsViewer As New frmRestaurantviewer
    frmRestaurantsViewer.ArcMapApp = m_application
    frmRestaurantsViewer.Show()


        End Sub
    ```

9. Now we can use the application object in the form. Open the `frmRestaurantsViewer.vb` in code view and import the following libraries:

    ```
    Imports ESRI.ArcGIS.Geodatabase
    Imports ESRI.ArcGIS.DataSourcesGDB
    Imports System.Windows.Forms
    Imports ESRI.ArcGIS.ArcMapUI
    Imports ESRI.ArcGIS.Display
    ```

```
Imports ESRI.ArcGIS.Geometry


Public Class frmRestaurantviewer
```

In the `btnHighlight_Click` method, add the following code to access the display object using the `_application` object we just set in the form. Note that we called the `StartDrawing` and `FinishDrawing` methods, and the code between those two lines is the one we will be writing to draw on the screen:

```
Private Sub btnHighlight_Click(sender As Object, e As
EventArgs) Handles btnHighlight.Click


        Dim pDocument As IMxDocument = _application.Document


        Dim pScreenDisplay As IScreenDisplay = pDocument.
ActiveView.ScreenDisplay
pScreenDisplay.StartDrawing(pScreenDisplay.hDC, ESRI.ArcGIS.
Display.esriScreenCache.esriNoScreenCache)


pScreenDisplay.FinishDrawing()

        End Sub
```

10. We need to set a basic `MarkerSymbol` to draw with; this will be a plain black point. Write the following code:

```
Private Sub btnHighlight_Click(sender As Object, e As EventArgs)
Handles btnHighlight.Click


        Dim pDocument As IMxDocument = _application.Document


        Dim pScreenDisplay As IScreenDisplay = pDocument.
ActiveView.ScreenDisplay
pScreenDisplay.StartDrawing(pScreenDisplay.hDC, ESRI.ArcGIS.
Display.esriScreenCache.esriNoScreenCache)


        Dim pMarkerSymbol As ISymbol = New SimpleMarkerSymbol

pScreenDisplay.SetSymbol(pMarkerSymbol)

pScreenDisplay.FinishDrawing()

        End Sub
```

11. The venues we want to draw are points, so we will use the `DrawPoint` method in order to draw. However, we need to get the selected venue point geometry. To do that, we need to query the feature class and get the feature using the object ID which we have saved in the drop-down list. We will then use the shape property in the feature to get the geometry, which is the point in this case. This is explained in the following code:

```
Private Sub btnHighlight_Click(sender As Object, e As
EventArgs) Handles btnHighlight.Click

        Dim pDocument As IMxDocument = _application.Document

        Dim pScreenDisplay As IScreenDisplay = pDocument.
ActiveView.ScreenDisplay
pScreenDisplay.StartDrawing(pScreenDisplay.hDC, ESRI.ArcGIS.
Display.esriScreenCache.esriNoScreenCache)

        Dim pMarkerSymbol As ISymbol = New SimpleMarkerSymbol

pScreenDisplay.SetSymbol(pMarkerSymbol)

        Dim pSelectedVenueKeyValuePair As KeyValuePair(Of Integer,
String) = cmbVenue.SelectedItem

        Dim pFeatureClass As IFeatureClass =
GetVenuesFeatureClass()
        Dim pFeature As IFeature = pFeatureClass.GetFeature(pSelec
tedVenueKeyValuePair.Key)

pScreenDisplay.DrawPoint(pFeature.Shape)
pScreenDisplay.FinishDrawing()

        End Sub
```

12. Build your solution and run the `bestaurants`.mxd.

13. Click on the **Bestaurants Viewer** button to view the **Bestaurants** form.

14. Select **Diner** from **Category** and then select the **Balan's Diner** venue.

15. Click on **Highlight**. Note that a black point is drawn on the diner, as illustrated in the following screenshot:

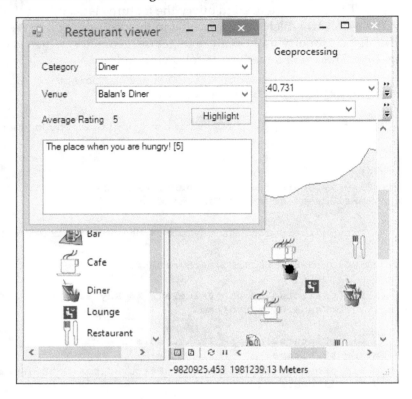

16. Try different venues; you will see that the black dots persist, however, when you refresh, they disappear.

17. Close ArcMap and choose not to save any changes.

Highlighting features

Now that we have learned how to use the display object to draw on the screen, we will use this option to make the venue blink, flash, or just highlight. This will involve drawing and erasing. To do that, we have to use a special pen object that allows us to draw over the point to erase it, which gives the feeling of highlighting. Follow these steps:

1. If necessary, open Visual Studio Express in Administrator mode; we need to do this since our project is actually writing to the registry this time, so it needs administrator permissions. To do that, right-click on Visual Studio and click on **Run as administrator**.

2. Go to **File | Open Project**, browse to the `Bestaurants` project from the `C:\ArcGISByExample\bestaurants\Code`, and click on **Open**.

3. Right-click on `frmRestaurantsViewer.vb` and select **View Code**.

4. In `btnHighlight_Click`, we will make a small change to our symbol to support erasing, as follows:

```
    Private Sub btnHighlight_Click(sender As Object, e As EventArgs)
Handles btnHighlight.Click

        Dim pDocument As IMxDocument = _application.Document

        Dim pScreenDisplay As IScreenDisplay = pDocument.
ActiveView.ScreenDisplay
pScreenDisplay.StartDrawing(pScreenDisplay.hDC, ESRI.ArcGIS.
Display.esriScreenCache.esriNoScreenCache)

        Dim pMarkerSymbol As ISymbol = New SimpleMarkerSymbol
        Dim pSymbol As ISymbol = pMarkerSymbol
pSymbol.ROP2 = esriRasterOpCode.esriROPNotXOrPen
pScreenDisplay.SetSymbol(pMarkerSymbol)

        Dim pSelectedVenueKeyValuePair As KeyValuePair(Of Integer,
String) = cmbVenue.SelectedItem

        Dim pFeatureClass As IFeatureClass =
GetVenuesFeatureClass()
        Dim pFeature As IFeature = pFeatureClass.GetFeature(pSelec
tedVenueKeyValuePair.Key)

pScreenDisplay.DrawPoint(pFeature.Shape)

pScreenDisplay.FinishDrawing()

    End Sub
```

5. Now we have to draw point and pause our code for few milliseconds and redraw again four times, which will give the appearance of flashing. Add the `DrawPoint` and `Sleep` methods as follows:

```
Private Sub btnHighlight_Click(sender As Object, e As
EventArgs) Handles btnHighlight.Click

        Dim pDocument As IMxDocument = _application.Document

        Dim pScreenDisplay As IScreenDisplay = pDocument.
ActiveView.ScreenDisplay
pScreenDisplay.StartDrawing(pScreenDisplay.hDC, ESRI.ArcGIS.
Display.esriScreenCache.esriNoScreenCache)

        Dim pMarkerSymbol As ISymbol = New SimpleMarkerSymbol
        Dim pSymbol As ISymbol = pMarkerSymbol
pSymbol.ROP2 = esriRasterOpCode.esriROPNotXOrPen
pScreenDisplay.SetSymbol(pMarkerSymbol)

        Dim pSelectedVenueKeyValuePair As KeyValuePair(Of Integer,
String) = cmbVenue.SelectedItem

        Dim pFeatureClass As IFeatureClass =
GetVenuesFeatureClass()
        Dim pFeature As IFeature = pFeatureClass.GetFeature(pSelec
tedVenueKeyValuePair.Key)

pScreenDisplay.DrawPoint(pFeature.Shape)
Threading.Thread.Sleep(100)
pScreenDisplay.DrawPoint(pFeature.Shape)
Threading.Thread.Sleep(100)
pScreenDisplay.DrawPoint(pFeature.Shape)
Threading.Thread.Sleep(100)
pScreenDisplay.DrawPoint(pFeature.Shape)

pScreenDisplay.FinishDrawing()

        End Sub
```

6. You can control the sleep period: the shorter the interval, the faster the blinking. Build your solution and run `bestaurants.mxd`.

7. Click on the **Bestaurants Viewer** button to view the **Bestaurants** form.

8. Select **Diner** from **Category** and then select the **Balan's Diner** venue.

9. Click on **Highlight** and you will see that the venue starts to blink four times with the black dot that we drew.

10. Close ArcMap and choose not to save any changes.

Filtering restaurants on the map

This is the last section in this chapter. We learn how to filter the map to display only desired venues. We will apply this filter on the category selection, for instance, when I select **Cafe's**, only cafes will be displayed on the map, and so on. Follow these steps to do so:

1. If necessary, open Visual Studio Express in Administrator mode; we need to do this since our project is actually writing to the registry this time, so it needs administrator permissions. To do that, right-click on Visual Studio and click on **Run as administrator**.

2. Go to **File | Open Project**, browse to the **Bestaurants** project from the `C:\ArcGISByExample\bestaurants\Code`, and click on **Open**.

3. Right-click on `frmRestaurantsViewer.vb` and select **View Code**.

4. In the `cmbCategory_SelectedIndexChanged` method, scroll to the end of the method and add the following code to get the layer and set the definition query of the layer, so that we only display the currently selected category:

```
    Private Sub cmbCategory_SelectedIndexChanged(sender As Object, e
As EventArgs) Handles cmbCategory.SelectedIndexChanged
        Dim pKeyValuePair As KeyValuePair(Of Integer, String) =
cmbCategory.SelectedItem

        Dim pFeatureClass As IFeatureClass =
GetVenuesFeatureClass()
```

```
        Dim pQFilter As IQueryFilter = New QueryFilter
pQFilter.WhereClause = "CATEGORY = " &pKeyValuePair.Key

        Dim pFeatureCursor As IFeatureCursor = pFeatureClass.
Search(pQFilter, False)

        Dim pFeature As IFeature = pFeatureCursor.NextFeature
        Dim dictVenues As New Dictionary(Of Integer, String)

        Do UntilpFeature Is Nothing
            Dim sName As String = pFeature.Value(pFeature.Fields.
FindField("NAME"))
            Dim oid As Integer = pFeature.OID
dictVenues.Add(oid, sName)
pFeature = pFeatureCursor.NextFeature
        Loop


cmbVenue.DisplayMember = "Value"
cmbVenue.ValueMember = "Key"
cmbVenue.DataSource = New BindingSource(dictVenues, Nothing)

        Dim pDocument As IMxDocument = _application.Document
        Dim pFeatureLayer As IFeatureLayerDefinition = pDocument.
FocusMap.Layer(0)
pFeatureLayer.DefinitionExpression = "CATEGORY = " &pKeyValuePair.
Key
pDocument.UpdateContents()
pDocument.ActiveView.Refresh()

    End Sub
```

5. Build your solution and run bestaurants.mxd.

6. Click on the **Bestaurants Viewer** button to view the **Bestaurants** form.

7. Select **Cafe** from **Category** and notice how the map refreshes to only show cafes, as illustrated in the following screenshot:

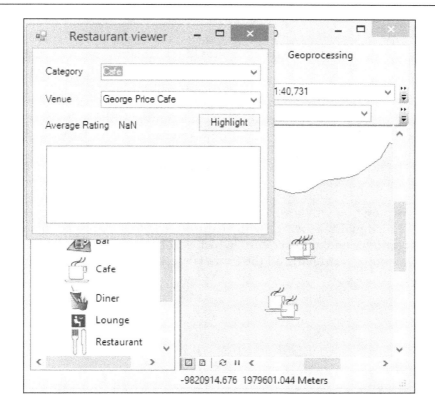

8. Try different categories to check that your code works fine.

9. Close ArcMap and choose not to save any changes.

Now you have learned how to filter the layers to only display certain features. You can use this in many other applications. You can find the latest source code under `B04847_06_Files\bestaurants\FinalCode`.

Summary

In this chapter, you added more functionality to the restaurant management application. In the first part of the chapter, you spent time learning about relationships in general and how the reviews and ratings table is related to the main restaurants feature class. Then you learned how to query these related tables, fetch vital data, and process these data to display useful information on the form. Then you learned about the display object and how you can use it to draw on the screen. You used this object to highlight the restaurants. Finally, you learned how to use the layer definition to filter the features on the map to only display certain restaurants based on a query.

In the next chapter, you will keep enhancing the restaurants management application to add more functionality to it. You will be introduced to advanced and real-time searching techniques and spatial query, which will help with writing advanced queries. For example, retrieving all the four-star restaurants in a particular region.

Advanced Searching

7

In the previous chapter, you learned about relationship queries in the `Restaurants` project. You learned how to display the ratings and the reviews by querying-related tables. In this chapter, we will learn some advanced query and search techniques. We will introduce a new geodatabase that contains the regions of Belize. We will add a quick search control at the toolbar so that we can find restaurants quickly in a given region.

In this chapter, we will discuss the following topics:

- Querying the regions
- Finding restaurants in a region
- Adding the search textbox to the toolbar
- Real-time search and filtering

Querying the regions

When dealing with real projects, often your data is not located in single database. You might need to connect to multiple databases or import those pieces of data into a single database in order to access. In this section, we will bring a second geodatabase, which contains the regions, connect to it from code, and retrieve the results to populate it in our form.

Connecting to the region's geodatabase

The regions feature class is located in another geodatabase that we have just added in this chapter. To prepare that feature class for access, follow these steps:

1. Copy the entire `bestaurants` folder in the supporting files for this chapter, `B04847_07_Files\bestaurants\` to the `C:\ArcGISByExample\`.

2. From the Start menu, run **ArcCatalog**.

3. Connect to the `C:\ArcGISByExample\` folder and browse to `C:\ArcGISbyExample\Data\bestaurants\Belize_Regions.gdb`.

4. Preview the `Regions` feature class in table view, as shown in the following screenshot. Note that we have seven regions named A to G. The region name is stored in the **NAME** field. This is important for the query that we will be making in the next section.

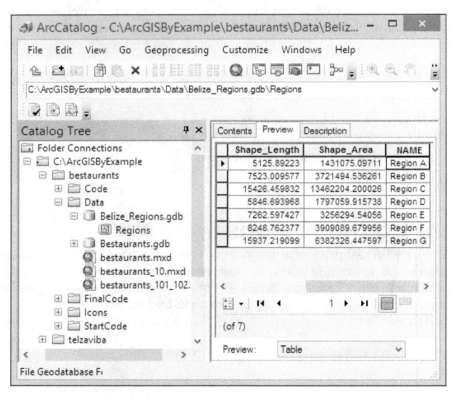

5. Close ArcCatalog.

6. Open Visual Studio Express in administrator mode; we need to do this since our project is actually writing to the registry this time, so it needs administrator permissions. To do that, right-click on Visual Studio and click on **Run as administrator**.

7. Pick up the starting code from `B04847_07_Files\bestaurants\StartCode` and copy it under `C:\ArcGISByExample\bestaurants\Code`. This is basically where we left off in *Chapter 6, Reviews and Ratings*.

8. Go to **File**, then click on **Open Project**, browse to the `Bestaurants` project from the `C:\ArcGISByExample\bestaurants\Code`, and click on **Open**.

9. Double-click on `frmRestaurantsViewer.vb` to open the form designer.

10. From the Toolbox, add a label, name it `lblRegion`, and add the Text property `Region`. Then, add a combo box, name it `cmbRegion`, which will have the list of regions, as illustrated in the following screenshot:

11. Now, open the code view of the form by right-clicking on `frmRestaurantsViewer.vb` and selecting **View Code**.

12. Write the `GetRegionsFeatureClass` function to connect to the regions geodatabase located in `C:\ArcGISByExample\bestaurants\Data\Belize_Regions.gdb` as follows:

```
Private Function GetRegionsFeatureClass() As IFeatureClass
        Dim pWorkspaceFactory As IWorkspaceFactory = New
FileGDBWorkspaceFactory
        Dim pWorkspace As IWorkspace = pWorkspaceFactory.
OpenFromFile("C:\ArcGISByExample\bestaurants\Data\Belize_Regions.
gdb", Me.Handle.ToInt32)

        Dim pFeatureWorkspace As IFeatureWorkspace = pWorkspace
        Dim pFeatureClass As IFeatureClass = pFeatureWorkspace.
OpenFeatureClass("Regions")
        Return pFeatureClass
    End Function
```

13. Save your code and move to the next section to query the regions and populate it in the list.

Populating the regions

Now that we have established a geodatabase connection to the regions feature class, we can query it and read the different regions in that table. Before we do this, we need to do some refactoring and grouping fragments of code into methods. Follow these steps to do so:

1. If necessary, open the `frmRestaurantsViewer.vb` form in code view.

2. The first change we will make is a small code refactoring, that is, move all the code in the `FrmRestaurantviewer_Load` method that populates **Categories** to a new method and call it `PopulateCategories`.

3. From the `FrmRestaurantviewer_Load` method, call the `PopulateCategories` method.

4. Write the `PopulateRegions` method that will use the regions feature class and loop through the features.

5. We will now refactor and create the `PopulateVenues` method. In the `cmbCategory_SelectedIndexChanged` method, move all the code after the line `Dim pKeyValuePair As KeyValuePair(Of Integer, String) = cmbCategory.SelectedItem` to a new method and call it `PopulateVenues` that accepts a parameter `categorycode`.

6. In the `PopulateVenues` method, change `pKeyValuePair.Key` to `categorycode` as follows:

```
Private Sub PopulateVenues(categorycode As Integer)

        Dim pFeatureClass As IFeatureClass =
GetVenuesFeatureClass()

        Dim pQFilter As IQueryFilter = New QueryFilter
        pQFilter.WhereClause = "CATEGORY = " &categorycode

        Dim pFeatureCursor As IFeatureCursor = pFeatureClass.
Search(pQFilter, False)

        Dim pFeature As IFeature = pFeatureCursor.NextFeature
        Dim dictVenues As New Dictionary(Of Integer, String)

        Do Until pFeature Is Nothing
            Dim sName As String = pFeature.Value(pFeature.Fields.
FindField("NAME"))
            Dim oid As Integer = pFeature.OID
            dictVenues.Add(oid, sName)
            pFeature = pFeatureCursor.NextFeature
        Loop


        cmbVenue.DisplayMember = "Value"
        cmbVenue.ValueMember = "Key"
        cmbVenue.DataSource = New BindingSource(dictVenues,
Nothing)

        Dim pDocument As IMxDocument = _application.Document
        Dim pFeatureLayer As IFeatureLayerDefinition = pDocument.
FocusMap.Layer(0)
        pFeatureLayer.DefinitionExpression = "CATEGORY = "
&categorycode
        pDocument.UpdateContents()
        pDocument.ActiveView.Refresh()
    End Sub
```

7. In the `cmbCategory_SelectedIndexChanged` method, call the `PopulateVenues` and pass the `pKeyValuePair.Key` value that indicates the region code, as shown in the following code:

```
Private Sub cmbCategory_SelectedIndexChanged(sender As Object,
e As EventArgs) Handles cmbCategory.SelectedIndexChanged
        Dim pKeyValuePair As KeyValuePair(Of Integer, String) =
cmbCategory.SelectedItem

PopulateVenues(pKeyValuePair.Key)
    End Sub
```

8. So far we didn't make any changes in the functionality of the application and it should work as usual. Optionally, build and test that the application works and that you didn't break anything.

9. It is time to write our `PopulateRegions` method; it should be similar in structure to `PopulateVenues`, except that we don't really need to pass any parameters for now. We have to connect to the regions feature class; note that we don't need to add a query filter since we need to get all regions:

```
Private Sub PopulateRegions()
        Dim pFeatureClass As IFeatureClass =
GetRegionsFeatureClass()

        Dim pFeatureCursor As IFeatureCursor = pFeatureClass.
Search(Nothing, False)

        Dim pFeature As IFeature = pFeatureCursor.NextFeature
        Dim dictRegions As New Dictionary(Of Integer, String)

        Do Until pFeature Is Nothing
            Dim sName As String = pFeature.Value(pFeature.Fields.
FindField("NAME"))
            Dim oid As Integer = pFeature.OID
            dictRegions.Add(oid, sName)
            pFeature = pFeatureCursor.NextFeature
        Loop


        cmbRegion.DisplayMember = "Value"
        cmbRegion.ValueMember = "Key"
        cmbRegion.DataSource = New BindingSource(dictRegions,
Nothing)

    End Sub
```

10. In the `FrmRestaurantviewer_Load` method, call the `PopulateRegions` method to populate the regions in the drop-down list when the form loads:

```
    Private Sub frmRestaurantviewer_Load(sender As Object, e As
EventArgs) Handles MyBase.Load
PopulateRegions()
        PopulateCategories()
    End Sub
```

11. Build your solution and run `bestaurants.mxd` in `C:\ArcGISByExample\Data\bestaurants.mxd`.

12. Click on the **Bestaurants viewer** button to view the **Bestaurants** form.

13. Note that the list of regions are now populated in the drop-down list; expand the list to see all seven regions, as illustrated in the following screenshot:

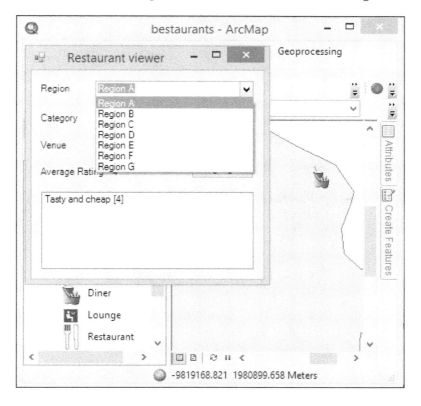

Note that we have the list of regions in this list and we will use this list to filter and find restaurants in a particular region. However, there are cases where you want to search restaurants in all regions. So this means we need to add one more entry in the regions drop-down list to signify **All Regions**.

14. Close ArcMap and choose not to save any changes.

15. Open `frmRestaurantsViewer.vb` in **Code View**.

16. In the `PopulateRegions` method, add the following line of code to add a new entry with key of `-1`, which we will make a special case for while filter and value of **All Regions**:

```
Private Sub PopulateRegions()
    Dim pFeatureClass As IFeatureClass =
GetRegionsFeatureClass()

    Dim pFeatureCursor As IFeatureCursor = pFeatureClass.
Search(Nothing, False)

    Dim pFeature As IFeature = pFeatureCursor.NextFeature
    Dim dictRegions As New Dictionary(Of Integer, String)

    Do Until pFeature Is Nothing
        Dim sName As String = pFeature.Value(pFeature.Fields.
FindField("NAME"))
        Dim oid As Integer = pFeature.OID
        dictRegions.Add(oid, sName)
        pFeature = pFeatureCursor.NextFeature
    Loop

    dictRegions.Add(-1, "All Regions")

    cmbRegion.DisplayMember = "Value"
    cmbRegion.ValueMember = "Key"
    cmbRegion.DataSource = New BindingSource(dictRegions,
Nothing)

    End Sub
```

17. Build and run your code; you should see the **All Regions** entry as shown in the following screenshot:

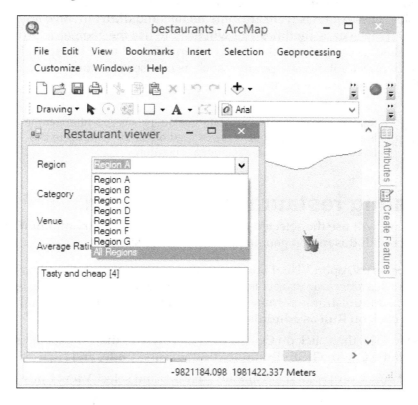

18. Close ArcMap and choose not to save any changes.

Finding restaurants in a region

We learned how to query the regions that were located in another geodatabase and populate them in a list. So far we didn't add any logic to filter restaurants in a particular region; here is where we can achieve that.

Spatial queries

In the previous six chapters, we learned how to query feature classes and even related tables. These queries were attribute related and didn't involve any spatial components. However, sometimes you will need to use the geometric property of the feature class in your query, and that is when you will need the spatial query. The **spatial query** uses a relational operator, such as intersects or contains, to retrieve all records that satisfy this criterion.

> Spatial query: This is a geodatabase query where a filter geometry is supplied and the query is executed on all the features in the feature class based on a relational operator.

Populating restaurants in a region

In our case, we will use the selected region as the filter geometry and find all venues that intersect with this region geometry. Follow these steps:

1. If necessary, open Visual Studio Express in administrator mode; we need to do this since our project is actually writing to the registry this time, so it needs administrator permissions. To do that, right-click on Visual Studio and click on **Run as administrator**.

2. Go to **File**, then click on **Open Project**, browse to the Bestaurants project from the **C:\ArcGISByExample\bestaurants\Code**, and click on **Open**.

3. Right-click on frmRestaurantsViewer.vb and select **View Code**.

4. Go to the PopulateVenues method and add one extra parameter besides categorycode, you guessed it regionoid. This is the object ID of the region:

```
    Private Sub PopulateVenues(categorycode As Integer, regionoid
As Integer)

        Dim pFeatureClass As IFeatureClass =
GetVenuesFeatureClass()
    ...
```

5. If you remember, we have created a query filter to search our feature class; this time we will use a better filter that does both spatial and attribute querying. The spatial filter, change the following code accordingly:

```
    Private Sub PopulateVenues(categorycode As Integer, regionoid
As Integer)
```

```
        Dim pFeatureClass As IFeatureClass =
GetVenuesFeatureClass()

        Dim pSFilter As ISpatialFilter = New SpatialFilter
        pSFilter.WhereClause = "CATEGORY = " & categorycode

        Dim pFeatureCursor As IFeatureCursor = pFeatureClass.
Search(pSFilter, False)

        Dim pFeature As IFeature = pFeatureCursor.NextFeature
        Dim dictVenues As New Dictionary(Of Integer, String)
    ...
```

6. Now we will add the filter geometry to the spatial filter; this is actually the selected region geometry. To get the selected region geometry, we need to get the feature, and to get the feature, we require the object ID and the region feature class; we have all that:

```
    Private Sub PopulateVenues(categorycode As Integer, regionoid
As Integer)

        Dim pFeatureClass As IFeatureClass =
GetVenuesFeatureClass()

        Dim pSelectedRegion As IFeature
        Dim pRegionFeatureClass As IFeatureClass =
GetRegionsFeatureClass()
        pSelectedRegion = pRegionFeatureClass.
GetFeature(regionoid)

        Dim pSFilter As ISpatialFilter = New SpatialFilter
        pSFilter.WhereClause = "CATEGORY = " & categorycode
        pSFilter.Geometry = pSelectedRegion.Shape
        pSFilter.SpatialRel = esriSpatialRelEnum.
esriSpatialRelIntersects

        Dim pFeatureCursor As IFeatureCursor = pFeatureClass.
Search(pSFilter, False)
    ...
    ...
```

7. Finally, we need to pass the region object ID to the `PopulateVenues` method, and add the following code in the `cmbCategory_SelectedIndexChanged`:

```
Private Sub cmbCategory_SelectedIndexChanged(sender As Object,
e As EventArgs) Handles cmbCategory.SelectedIndexChanged
    Dim pKeyValuePair As KeyValuePair(Of Integer, String) =
cmbCategory.SelectedItem

    Dim pKeyValueRegion As KeyValuePair(Of Integer, String) =
cmbRegion.SelectedItem

    PopulateVenues(pKeyValuePair.Key, pKeyValueRegion.Key)
End Sub
```

8. Go to **Build** and then click on **Build Solution**. Make sure ArcMap is not running. If you got an error, make sure you have run the Visual Studio as administrator.

9. Run `bestaurants.mxd` under `C:\ArcGISByExample\bestaurants\Data\ bestaurants.mxd`.

10. Click on the **Bestaurants viewer** button to view the **Bestaurants** form.

11. Select **Region F** from **Region** and then select **Diner** from the **Category** venue. You should see only two diners, as shown in the following screenshot:

12. Confirm that this is true by adding the **Regions** feature class to your Bestaurants map document, as shown in the next screenshot:

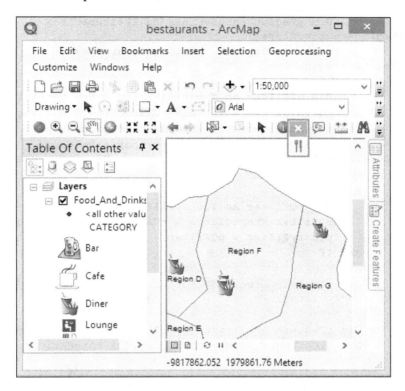

13. Close ArcMap and choose not to save any changes.

14. We still didn't support the **All Regions** selection. If you select this with our current code, you will definitely get an error since the object ID for that particular entry is -1 and there is no such object ID in the geodatabase. To fix that, we need to add an `if` condition in order to ignore the spatial query when we select all regions. This is basically our original case where we display all venues in a category. This is illustrated as follows in PopulateVenues:

```
Private Sub PopulateVenues(categorycode As Integer, regionoid As
Integer)

        Dim pFeatureClass As IFeatureClass =
GetVenuesFeatureClass()
```

```
        Dim pQueryFilter As IQueryFilter
        If regionoid <> -1 Then
            Dim pSelectedRegion As IFeature
            Dim pRegionFeatureClass As IFeatureClass =
GetRegionsFeatureClass()
            pSelectedRegion = pRegionFeatureClass.
GetFeature(regionoid)

            Dim pSFilter As ISpatialFilter = New SpatialFilter
            pSFilter.WhereClause = "CATEGORY = " & categorycode
            pSFilter.Geometry = pSelectedRegion.Shape
            pSFilter.SpatialRel = esriSpatialRelEnum.
esriSpatialRelIntersects
            pQueryFilter = pSFilter
        Else
            Dim pQFilter As IQueryFilter = New QueryFilter
            pQFilter.WhereClause = "CATEGORY = " & categorycode
            pQueryFilter = pQFilter
        End If

        Dim pFeatureCursor As IFeatureCursor = pFeatureClass.
Search(pQueryFilter, False)

        Dim pFeature As IFeature = pFeatureCursor.NextFeature
        Dim dictVenues As New Dictionary(Of Integer, String)
''
```

15. Go to **Build** and then click on **Build Solution**. Make sure ArcMap is not running. If you got an error, make sure you have run the Visual Studio as administrator.

16. Run `bestaurants.mxd` under `C:\ArcGISByExample\bestaurants\Data\bestaurants.mxd`.

17. Click on the **Bestaurants viewer** button to view the **Bestaurants** form.

18. Select **All Regions** from **Region** and then select **Diner** from the **Category** venue. You should see all of the diners now, as shown in the following screenshot:

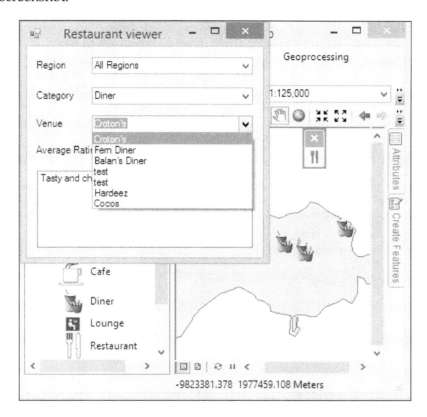

Sometimes the query might return no results. In that case, the drop-down list will be populated with an empty collection. You can handle this case and display something else instead, such as "No results."

Adding the search textbox in the toolbar

We have learned in the previous chapters how we can add a button and a tool to the toolbar. However, we can actually add more useful control to the toolbar that doesn't exist by default, by extending ArcObjects project in a similar way to a textbox or a dropdown menu. We can add almost any control in windows to the ArcGIS toolbar (as long as it fits of course). Here we will see how to add a textbox and use it to filter the map later. First we will add a new form, add a textbox to it, and resize the form to fit the text. Then we will add a button to the toolbar and replace it with the text control. Follow these steps to do so:

1. If necessary, open Visual Studio Express in administrator mode.

2. Go to **File**, then click on **Open Project**, browse to the `Bestaurants` project from the `C:\ArcGISByExample\bestaurants\Code`, and click on **Open**.

3. From the **Project** menu, click on **Add Windows Form** and name it `txtSearchForm.vb`.

4. In the form properties, scroll until you find **FormBorderStyle** and choose **None** to remove the bar of the form, as shown in the following screenshot:

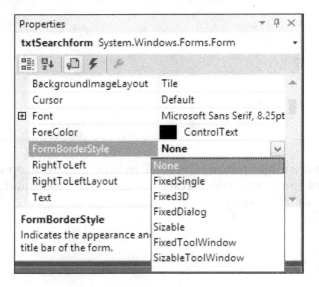

5. Add a textbox to the form from the toolbox and name it `txtSearch`.

6. In their respective properties, apply the width of `157` and height of `20` for both your `txtSearch` textbox and `txtSearchForm` form.

7. We have prepared our text control for ArcGIS time to add our ArcGIS button. Go to the **Project** menu and click on **Add Class**.

8. Expand **ArcGIS | Extending ArcObjects**, choose **Base Command**, name it cmSearch.vb, and then click on **Add**.

9. Choose **Desktop ArcMap Command** from **ArcGIS new Item Wizard Options** and then click on **OK**.

10. In the New method, write the following:

```
Public Sub New()
        MyBase.New()

        MyBase.m_category = "Bestaurants"  'localizable text
        MyBase.m_caption = "Bestaurants Quick Search"
        MyBase.m_message = "Bestaurants Quick Search"
        MyBase.m_toolTip = "Bestaurants Quick Search"
        MyBase.m_name = "Bestaurants_BestaurantsQuickSearch"

        Try
            'TODO: change bitmap name if necessary
            Dim bitmapResourceName As String = Me.GetType().Name +
".bmp"
            MyBase.m_bitmap = New Bitmap(Me.GetType(),
bitmapResourceName)
        Catch ex As Exception
            System.Diagnostics.Trace.WriteLine(ex.Message,
"Invalid Bitmap")
        End Try
    End Sub
```

11. We need to declare a modular instance of txtSearchForm in this command so that we can get the handle to it and add it to the toolbar. Add the following line right before Public Sub New():

```
Private m_application As IApplication
Private _txtsearchform As New txtSearchform

Public Sub New()
        MyBase.New()
    ...
```

12. Next we will implement the interface `IToolControl` to get access to the handle of the button and change it to our form. Write the following code after `Inherits BaseCommand` and press *Enter* to generate the implementation code:

```
<ComClass(cmSearch.ClassId, cmSearch.InterfaceId, cmSearch.
EventsId), _
 ProgId("Bestaurants.cmSearch")> _
Public NotInheritable Class cmSearch
    Inherits BaseCommand
Implements ESRI.ArcGIS.SystemUI.IToolControl
```

13. Three methods will be generated, `hWnd`, `onDrop`, and `onFocus`, but we are interested in `hWnd`. Write the following code in the `hWnd` method to return the `txtSearchForm` handle, this will trick Windows into showing our form instead of the button:

```
Public ReadOnly Property hWnd As Integer Implements ESRI.
ArcGIS.SystemUI.IToolControl.hWnd
        Get
            Return _txtsearchform.Handle.ToInt32
        End Get
    End Property
```

14. We are almost ready. We just need to add our button to the toolbar. Remember that we need to add to the toolbar items that ArcGIS understands, and ArcGIS does understand the `cmSearch` button. Double-click on the `tbBestaurants.vb` toolbar class to view the code. Add the following code to the `New` method:

```
Public Sub New()

AddItem("Bestaurants.cmSearch")
        AddItem("Bestaurants.cmViewer")

End Sub
```

15. Go to **Build** and then click on **Build Solution**. Make sure ArcMap is not running. If you got an error, make sure you have run the Visual Studio as administrator.

16. Run `bestaurants.mxd` under `C:\ArcGISByExample\bestaurants\Data\bestaurants.mxd`.

17. Take a look at how our new quick search looks like. Note that the text appeared before our Bestaurants viewer, and that is normal since we have added it before, as it shows in the sequence in the following screenshot:

18. Close ArcMap and choose not to save any changes.

Real-time search and filtering

After adding the textbox right on the toolbar, it will make it easier for us to search for restaurants. Let us code this control:

1. If necessary, open Visual Studio Express in administrator mode.

2. Go to **File**, then click on **Open Project**, browse to the Bestaurants project from the C:\ArcGISByExample\bestaurants\Code, and click on **Open**.

3. We need the txtSearchform control to have access to ArcMap application. For that, we need to add an Application property and then set it. Right-click on txtSearchForm.vb and select **View Code**.

4. Add the following property to the code:

```
Public Class txtSearchform

    Private _application As ESRI.ArcGIS.Framework.IApplication
    Public Property ArcMapApplication() As ESRI.ArcGIS.Framework.
IApplication
        Get
            Return _application
        End Get
        Set(ByVal value As ESRI.ArcGIS.Framework.IApplication)
            _application= value
        End Set
    End Property
End Class
```

5. Double-click on cmSearch.vb to view the code and add the following line to the onCreate method; this is where we set the application to the txtSearchForm:

```
Public Overrides Sub OnCreate(ByVal hook As Object)
    If Not hook Is Nothing Then
        m_application = CType(hook, IApplication)

        'Disable if it is not ArcMap
        If TypeOf hook Is IMxApplication Then
            MyBase.m_enabled = True
        Else
            MyBase.m_enabled = False
        End If
    End If
    ' TODO:  Add other initialization code
    _txtsearchform.ArcMapApplication = m_application

End Sub
```

6. Double-click on txtSearchForm to open the Form Designer.

7. Double-click on txtSearch to generate the txtSearch_TextChanged method. We want the search code to be executed as we type the restaurant name. Therefore, this is where the code should be written. If you decide that the search should be executed when the user hit the *Enter* key, then you can use the KeyPress method instead.

8. Use the application object to retrieve the `Food_and_Drinks` layer and then set the definition filter. Note that we want to search by the restaurant name. Write the following code:

```
Private Sub txtSearch_TextChanged(sender As Object, e As
EventArgs) Handles txtSearch.TextChanged

    Dim pDocument As ESRI.ArcGIS.ArcMapUI.IMxDocument = _
application.Document
    Dim pFeatureLayer As ESRI.ArcGIS.Carto.
IFeatureLayerDefinition = pDocument.FocusMap.Layer(0)
    pFeatureLayer.DefinitionExpression = "NAME = '" &
txtSearch.Text & "'"
    pDocument.UpdateContents()
    pDocument.ActiveView.Refresh()

End Sub
```

9. This code will search the name of the restaurant in a case sensitive manner, plus it will look for matching records. We need to use the database `LIKE` keyword and `UPPER` as shown in the following code. This will give us better results:

```
Private Sub txtSearch_TextChanged(sender As Object, e As
EventArgs) Handles txtSearch.TextChanged

    Dim pDocument As ESRI.ArcGIS.ArcMapUI.IMxDocument = _
application.Document
    Dim pFeatureLayer As ESRI.ArcGIS.Carto.
IFeatureLayerDefinition = pDocument.FocusMap.Layer(0)
    pFeatureLayer.DefinitionExpression = "UPPER(NAME) LIKE '%"
& txtSearch.Text.ToUpper & "%'"
    pDocument.UpdateContents()
    pDocument.ActiveView.Refresh()

End Sub
```

10. Go to **Build** and then click on **Build Solution**. Make sure ArcMap is not running.

11. Run `bestaurants.mxd` under `C:\ArcGISByExample\bestaurants\Data\bestaurants.mxd`.

12. Type `starbucks` in the quick search box and see how the map starts filtering the venues as you type in, as shown in the following screenshot. Use the identify tool in ArcMap to make sure your search is accurate.

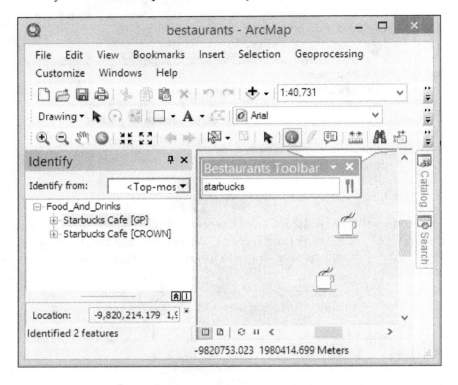

13. Close ArcMap and choose not to save any changes.

You have successfully completed the `bestaurant` application. You can find the latest source code under `B04847_07_Files\bestaurants\FinalCode`.

Summary

In this chapter, you completed the restaurants management application. In the first part of the chapter, you connected to another geodatabase to retrieve the regions. You then used the regions to intersect with venues using the power of spatial queries. With this you were able to find different restaurants in a given region and also identify which regions lack restaurants. Finally, you learned more ArcObjects techniques and how to harness the windows control and bind them with the ArcGIS controls. This is where you added the textbox on the toolbar and used it to search for restaurants in real time.

In the next chapter, you will start working on the final example, which is the excavation manager of Yahrnam. The excavation planning manager helps construction workers plan their excavation for utilities and telecom networks beforehand in a given area and estimate the cost of excavation. You will be introduced to geodatabase editing and more advanced ArcObjects techniques.

8
App 3 – Advanced ArcObjects

In the last seven chapters, we worked on cell tower analysis tool that allowed us to introduce ArcGIS and ArcGIS add-ins. We have created the bestaurants application that allowed us to manage and search restaurants introducing us to the extending ArcObjects approach. We learned so much about ArcGIS development, which makes us ready to take it to the next level. This chapter and the next two will be dedicated to the final challenging example, the excavation planning manager.

In this chapter, we will discuss the following topics:

- Geodatabase editing
- Preparing the data and project
- Creating excavation features
- Viewing and editing excavation information

Geodatabase editing

YharanamCo is a construction contractor experienced in executing efficient and economical excavations for utility and telecom companies. When YharanamCo's board of directors heard of ArcGIS technology, they wanted to use their expertise with the power of ArcGIS to come up with a solution that helps them cut costs even more. Soil type is not the only factor in the excavation, there are many factors including the green factor where you need to preserve the trees and green area while excavating for visual appeal. Using ArcGIS, YharanamCo can determine the soil type and green factor and calculate the cost of an excavation.

The excavation planning manager is the application you will be writing on top of ArcGIS. This application will help YharanamCo to create multiple designs and scenarios for a given excavation. This way they can compare the cost for each one and consider how many trees they could save by going through another excavation route. YharanamCo has provided us with the geodatabase of a soil and trees data for one of their new projects for our development.

So far we learned how to view and query the geodatabase and we were able to achieve that by opening what we called a workspace. However, changing the underlying data requires establishing an editing session. All edits that are performed during an edit sessions are queued, and the moment the session is saved, these edits are committed to the geodatabase. Geodatabase editing supports atomic transactions, which are referred to as operations in the geodatabase.

 Atomic transaction is a list of database operations that either all occur together or none. This is to ensure consistency and integrity.

After this short introduction to geodatabase editing, we will prepare our data and project.

Preparing the data and project

Before we dive into the coding part, we need to do some preparation for our new project and our data.

Preparing the Yharnam geodatabase and map

The YharnamCo team has provided us with the geodatabase and map document, so we will simply copy the necessary files to your drive. Follow these steps to start your preparation of the data and map:

1. Copy the entire yharnam folder in the supporting files for this chapter B04847_08_Files\yharnam\ to C:\ArcGISByExample\.

2. Run yharnam.mxd under C:\ArcGISByExample\yharnam\Data\yharnam. mxd. This should point to the geodatabase, which is located under C:\ ArcGISByExample\yharnam\Data\yharnam.gdb, as illustrated in the following screenshot:

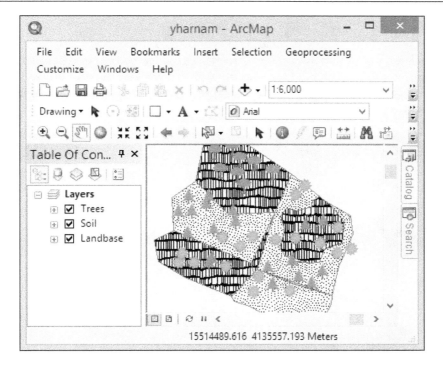

3. Note there are three types of trees: Type1, Type2, and Type3. Also note there are two types of soil: rocky and sand. This will be crucial in the next chapter in which we will calculate the cost of removing a tree.

4. Close ArcMap and choose not to save any changes.

Preparing the Yharnam project

We will now start our project. First we need to create our Yharnam Visual Studio extending ArcObjects project. To do so, follow these steps:

1. From the Start menu, run Visual Studio Express 2013 as administrator.

2. Go to the **File** menu and then click on **New Project**.

3. Expand the **Templates** node | **Visual Basic** | **ArcGIS**, and then click on **Extending ArcObjects**. You will see the list of projects displayed on the right.

4. Select the Class Library (ArcMap) project.

5. In the **Name** field, type Yharnam, and in the location, browse to C:\ArcGISByExample\yharnam\Code. If the Code folder is not there, create it.

6. Click on **OK**.

7. In the ArcGIS Project Wizard, you will be asked to select the references libraries you will need in your project. I always recommend selecting all referencing, and then at the end of your project, remove the unused ones. So, go ahead and right-click on **Desktop ArcMap** and click on **Select All**, as shown in the following screenshot:

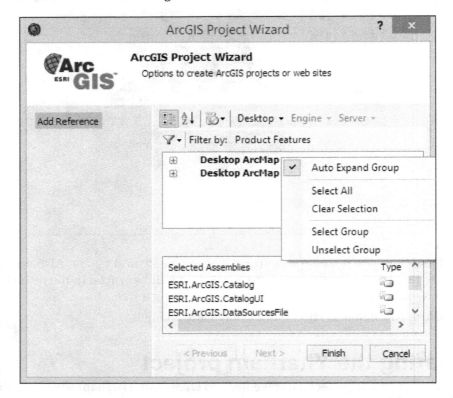

8. Click on **Finish** to create the project. This will take a while to add all references to your project.

9. Once your project is created, you will see that one class called Class1 is added, which we won't need, so right-click on it and choose **Delete**. Then, click on **OK** to confirm.

10. Go to **File** and click on **Save All**.

11. Exit the Visual Studio application.

You have finished preparing your Visual Studio with extending ArcObjects support. Move to the next section to write some code.

Adding the new excavation tool

The new excavation tool will be used to draw a polygon on the map, which represents the geometry of the excavation. Then, this will create a corresponding excavation feature using this geometry:

1. If necessary, open Visual Studio Express in administrator mode; we need to do this since our project is actually writing to the registry this time, so it needs administrator permissions. To do that, right-click on **Visual Studio** and click on Run as administrator.

2. Go to **File**, then click on Open Project, browse to the Yharnam project from C:\ArcGISByExample\yharnam\Code, and click on **Open**.

3. Click on the Yharnam project from **Solution Explorer** to activate it.

4. From the **Project** menu, click on **Add Class**.

5. Expand the ArcGIS node and then click on the **Extending ArcObjects** node.

6. Select **Base Tool** and name it tlNewExcavation.vb.

7. Click on **Add** to open **ArcGIS New Item Wizard Options**.

8. From **ArcGIS New Item Wizard Options**, select **Desktop ArcMap** tool since we will be programming against ArcMap. Click on **OK**.

9. Take note of the Yharnam.tlNewExcavation Progid as we will be using this in the next section to add the tool to the toolbar.

10. If necessary, double-click on tlNewExcavation.vb to edit it.

11. In the New method, update the properties of the command as follows. This will update the name and caption and other properties of the command. There is a piece of code that loads the command icon. Leave that unchanged:

```
Public Sub New()
MyBase.New()

' TODO: Define values for the public properties
' TODO: Define values for the public properties
MyBase.m_category = "Yharnam"  'localizable text
MyBase.m_caption = "New Excavation"   'localizable text
MyBase.m_message = "New Excavation"   'localizable text
MyBase.m_toolTip = "New Excavation" 'localizable text
MyBase.m_name = "Yharnam_NewExcavation"

        Try
            'TODO: change resource name if necessary
            Dim bitmapResourceName As String = Me.GetType().Name +
            ".bmp"
```

```
MyBase.m_bitmap = New Bitmap(Me.GetType(), bitmapResourceName)
MyBase.m_cursor = New System.Windows.Forms.Cursor(Me.GetType(),
Me.GetType().Name + ".cur")
        Catch ex As Exception
System.Diagnostics.Trace.WriteLine(ex.Message, "Invalid Bitmap")
        End Try
    End Sub
```

Adding the excavation editor tool

The excavation editor is a tool that will let us click an excavation feature on the map and display the excavation information such as depth, area, and so on. It will also allow us to edit some of the information.

We will now add a tool to our project. To do that, follow these steps:

1. If necessary, open Visual Studio Express in administrator mode; we need to do this since our project is actually writing to the registry this time, so it needs administrator permissions. To do that, right-click on **Visual Studio** and click on **Run as administrator**.

2. Go to **File**, then click on **Open Project**, browse to the `Yharnam` project from `C:\ArcGISByExample\yharnam\Code`, and click on **Open**.

3. Click on the `Yharnam` project from **Solution Explorer** to activate it.

4. From the **Project** menu, click on **Add Class**.

5. Expand the **ArcGIS** node and then click on the **Extending ArcObjects** node.

6. Select **Base Tool** and name it `tlExcavationEditor.vb`.

7. Click on **Add** to open **ArcGIS New Item Wizard Options**.

8. From **ArcGIS New Item Wizard Options**, select **Desktop ArcMap Tool** since we will be programming against ArcMap. Click on **OK**.

9. Take note of the `Yharnam.tlExcavationEditor` Progid as we will be using this in the next section to add the tool to the toolbar.

10. If necessary, double-click on `tlExcavationEditor.vb` to edit it.

11. In the `New` method, update the properties of the command as follows. This will update the name and caption and other properties of the command:
    ```
    MyBase.m_category = "Yharnam" 'localizable text
    MyBase.m_caption = "Excavation Editor"   'localizable text
    MyBase.m_message = "Excavation Editor"   'localizable text
    MyBase.m_toolTip = "Excavation Editor" 'localizable text
    MyBase.m_name = "Yharnam_ExcavationEditor"
    ```

12. In order to display the excavation information, we will need a form. To add the **Yharnam Excavation Editor** form, point to **Project** and then click on **Add Windows Form**.

13. Name the form `frmExcavationEditor.vb` and click on **Add**.

14. Use the form designer to add and set the controls shown in the following table:

Control	Name	Properties
Label	lblDesignID	Text: Design ID
Label	lblExcavationOID	Text: Excavation ObjectID
Label	lblExcavationArea	Text: Excavation Area
Label	lblExcavationDepth	Text: Excavation Depth
Label	lblTreeCount	Text: Number of Trees
Label	lblTotalCost	Text: Total Excavation Cost
Text	txtDesignID	Read-Only: True
Text	txtExcavationOID	Read-Only: True
Text	txtExcavationArea	Read-Only: True
Text	txtExcavationDepth	Read-Only: False
Text	txtTreeCount	Read-Only: True
Text	txtTotalCost	Read-Only: True
Button	btnSave	Text: Save

15. Your form should look like the following screenshot:

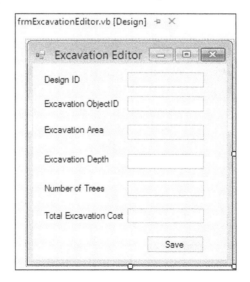

16. One last change before we build our solution: we need to change the default icons of our tools. To do that, double-click on `tlExcavationEditor.bmp` to open the picture editor and replace the picture with `yharnam.bmp`, which can be found under `C:\ArcGISByExample\yharnam\icons\excavation_editor.bmp`, and save `tlExcavationEditor.bmp`.

17. Change `tlNewExcavation.bmp` to `C:\ArcGISByExample\yharnam\icons\new_excavation.bmp`.

18. Save your project and move to the next step to assign the command to the toolbar.

Adding the excavation manager toolbar

Now that we have our two tools, we will add a toolbar to group them together. Follow these steps to add the Yharnam Excavation Planning Manager Toolbar to your project:

1. If necessary, open Visual Studio Express in administrator mode; we need to do this since our project is actually writing to the registry this time, so it needs administrator permissions. To do that, right-click on **Visual Studio** and click on **Run as administrator**.

2. Go to **File**, then click on **Open Project**, browse to the `Yharnam` project from `C:\ArcGISByExample\yharnam\Code`, and click on **Open**.

3. Click on the `Yharnam` project from **Solution Explorer** to activate it.

4. From the **Project** menu, click on **Add Class**.

5. Expand the **ArcGIS** node and then click on the **Extending ArcObjects** node.

6. Select **Base Toolbar** and name it `tbYharnam.vb`.

7. Click on **Add** to open **ArcGIS New Item Wizard Options**.

8. From **ArcGIS New Item Wizard Options**, select **Desktop ArcMap** since we will be programming against ArcMap. Click on **OK**.

9. The property **Caption** is what is displayed when the toolbar loads. It currently defaults to **MY VB.Net Toolbar**, so change it to **Yharnam Excavation Planning Manager Toolbar** as follows:

```
Public Overrides ReadOnly Property Caption() As String
        Get
                'TODO: Replace bar caption
                Return "YharnamExcavation Planning Manager"
        End Get
    End Property
```

10. Your toolbar is currently empty, which means it doesn't have buttons or tools. Go to the `New` method and add your tools prog ID, as shown in the following code:

```
    Public Sub New()
AddItem("Yharnam.tlNewExcavation")
AddItem("Yharnam.tlExcavationEditor")
    End Sub
```

11. Now it is time to test our new toolbar. Go to **Build** and then click on **Build Solution**; make sure ArcMap is not running. If you get an error, make sure you have run the Visual Studio as administrator.

 For a list of all ArcMap commands, you can refer to `http://bit.ly/b04748_arcmapids`. Check the commands with namespace `esriArcMapUI`.

12. Run `yharnam.mxd` in `C:\ArcGISByExample\yharnam\Data\yharnam.mxd`.

13. From **ArcMap**, go to the **Customize** menu, then to **Toolbars**, and then select **Yharnam Excavation Planning Manager Toolbar**, which we just created. You should see the toolbar pop up on ArcMap with the two added commands, as shown in the following screenshot:

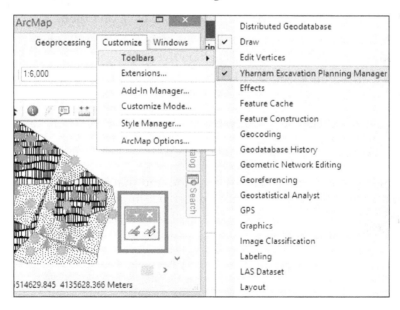

14. Close ArcMap and choose not to save any changes.

In the next section, we will do the real work of editing.

Creating excavation features

Features are nothing but records in a table. However, these special records cannot be created without a geometry shape attribute. To create a feature, we need first to learn how to draw and create geometries. We will be using the rubber band ArcObjects interface to create a polygon geometry. We can use it to create other types of geometries as well, but since our excavations are polygons, we will use the polygon rubber band.

Using the rubber band to draw geometries on the map

In this exercise, we will use the rubber band object to create a polygon geometry by clicking on multiple points on the map. We will import libraries as we need them. Follow these steps:

1. If necessary, open Visual Studio Express in administrator mode; we need to do this since our project is actually writing to the registry this time, so it needs administrator permissions. To do that, right-click on **Visual Studio** and click on **Run as administrator**.

2. Go to **File**, then click on **Open Project**, browse to the Yharnam project from C:\ArcGISByExample\yharnam\Code, and click on **Open**.

3. Double-click on tlNewExcavation.vb and write the following code in onMouseDown that is when we click on the map:

    ```
    Public Overrides Sub OnMouseDown(ByVal Button As Integer, ByVal
    Shift As Integer, ByVal X As Integer, ByVal Y As Integer)

        Dim pRubberBand As IRubberBand = New RubberPolygonClass

        End Sub
    ```

4. You will get an error under rubber band and that is because the library that this class is located in is not imported. Simply hover over the error and import the library; in this case, it is Esri.ArcGIS.Display, as illustrated in the following screenshot:

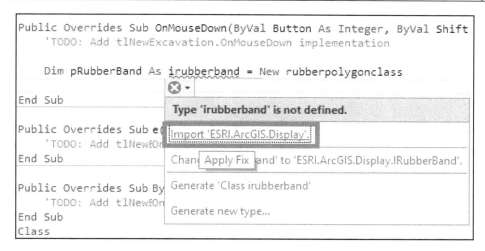

```
Public Overrides Sub OnMouseDown(ByVal Button As Integer, ByVal Shift
    'TODO: Add tlNewExcavation.OnMouseDown implementation

    Dim pRubberBand As irubberband = New rubberpolygonclass
```

5. We now have to call the `TrackNew` method on the `pRubberband` object that will allow us to draw. This requires two pieces of parameters. First, the screen on which you are drawing and the symbol you are drawing with, which have the color, size, and so on. By now we are familiar with how we can get these objects. The symbol needs to be of type `FillShapeSymbol` since we are dealing with polygons. We will go with a simple black symbol for starters. Write the following code:

```
Public Overrides Sub OnMouseDown(ByVal Button As Integer,
ByVal Shift As Integer, ByVal X As Integer, ByVal Y As Integer)

    Dim pDocument As IMxDocument = m_application.Document

    Dim pRubberBand As IRubberBand = New RubberPolygonClass

    Dim pFillSymbol As ISimpleFillSymbol = New
SimpleFillSymbol

Dim pPolygon as IGeometry=pRubberBand.TrackNew(pDocument.
ActiveView.ScreenDisplay, pFillSymbol)
    End Sub
```

6. Build your solution. If it fails, make sure you have run the solution as administrator.

7. Run `yharnam.mxd`.

8. Click on the **New Excavation** tool to activate it, and then click on three different locations on the map. You will see that a polygon is forming as you click; double-click to finish drawing, as illustrated in the following screenshot:

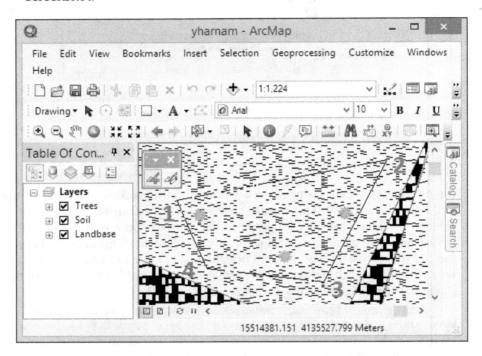

9. The polygon disappears when you finish drawing and the reason is that we didn't actually persist the polygon into a feature or a graphic. As a start, we will draw the polygon on the screen and we will also change the color of the polygon to red. Write the following code to do so:

```
Public Overrides Sub OnMouseDown(ByVal Button As Integer,
ByVal Shift As Integer, ByVal X As Integer, ByVal Y As Integer)

    Dim pDocument As IMxDocument = m_application.Document

    Dim pRubberBand As IRubberBand = New RubberPolygonClass

    Dim pFillSymbol As ISimpleFillSymbol = New
SimpleFillSymbol
    Dim pColor As IColor = New RgbColor
pColor.RGB = RGB(255, 0, 0)
pFillSymbol.Color = pColor
```

```
        Dim pPolygon As IGeometry = pRubberBand.
TrackNew(pDocument.ActiveView.ScreenDisplay, pFillSymbol)

        Dim pDisplay As IScreenDisplay = pDocument.ActiveView.
ScreenDisplay

pDisplay.StartDrawing(pDisplay.hDC, ESRI.ArcGIS.Display.
esriScreenCache.esriNoScreenCache)
pDisplay.SetSymbol(pFillSymbol)
pDisplay.DrawPolygon(pPolygon)
pDisplay.FinishDrawing()

    End Sub
```

10. Build your solution and run `yharnam.mxd`.

11. Activate the **New Excavation** tool and draw an excavation; you should see a red polygon is displayed on the screen after you finish drawing, as shown in the following screenshot:

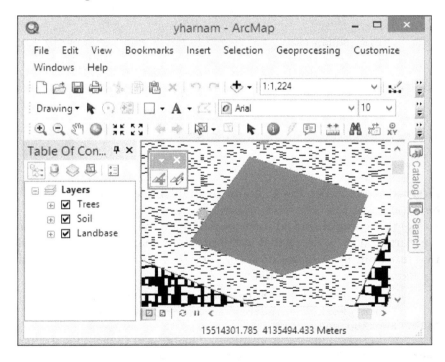

12. Close ArcMap and choose not to save the changes.

Converting geometries into features

Now that we have learned how to draw a polygon, we will convert that polygon into an excavation feature. Follow these steps to do so:

1. If necessary, open Visual Studio Express in administrator mode; we need to do this since our project is actually writing to the registry this time, so it needs administrator permissions. To do that, right-click on **Visual Studio** and click on **Run as administrator**.

2. Go to **File**, then click on **Open Project**, browse to the `Yharnam` project from `C:\ArcGISByExample\yharnam\Code`, and click on **Open**.

3. Double-click on `tlNewExcavation.vb` to edit the code.

4. Remove the code that draws the polygon on the map. Your new code should look like the following:

```
Public Overrides Sub OnMouseDown(ByVal Button As Integer, ByVal
Shift As Integer, ByVal X As Integer, ByVal Y As Integer)

        Dim pDocument As IMxDocument = m_application.Document

        Dim pRubberBand As IRubberBand = New RubberPolygonClass

        Dim pFillSymbol As ISimpleFillSymbol = New
SimpleFillSymbol

        Dim pPolygon As IGeometry = pRubberBand.
TrackNew(pDocument.ActiveView.ScreenDisplay, pFillSymbol)

    End Sub
```

5. First we need to open the `Yharnam` geodatabase located on `C:\ ArcGISByExample\yharnam\Data\Yharnam.gdb` by establishing a workspace connection and then we will open the `Excavation` feature class; we learned how to do all that in the previous chapters. Write the following two functions in `tlNewExcavation.vb`, `getYharnamWorkspace`, and `getExcavationFeatureclass`:

```
Public Function getYharnamWorkspace() As IWorkspace
        Dim pWorkspaceFactory As IWorkspaceFactory = New
FileGDBWorkspaceFactory
        Return pWorkspaceFactory.OpenFromFile("C:\ArcGISByExample\
yharnam\Data\Yharnam.gdb", m_application.hWnd)
    End Function
```

```
Public Function getExcavationFeatureClass(pWorkspace As
IWorkspace) As IFeatureClass
        Dim pFWorkspace As IFeatureWorkspace = pWorkspace
        Return pFWorkspace.OpenFeatureClass("Excavation")
    End Function
```

6. To create the feature, we need first to start an editing session and a transaction, and wrap and code between the start and the end of the session. To use editing, we utilize the IWorkspaceEdit interface. Write the following code in the onMouseDown method:

```
Public Overrides Sub OnMouseDown(ByVal Button As Integer,
ByVal Shift As Integer, ByVal X As Integer, ByVal Y As Integer)

        Dim pDocument As IMxDocument = m_application.Document

        Dim pRubberBand As IRubberBand = New RubberPolygonClass

        Dim pFillSymbol As ISimpleFillSymbol = New
SimpleFillSymbol

        Dim pPolygon As IGeometry = pRubberBand.
TrackNew(pDocument.ActiveView.ScreenDisplay, pFillSymbol)

        Dim pWorkspaceEdit As IWorkspaceEdit =
getYharnamWorkspace()
pWorkspaceEdit.StartEditing(True)
pWorkspaceEdit.StartEditOperation()


pWorkspaceEdit.StopEditOperation()
pWorkspaceEdit.StopEditing(True)

    End Sub
```

7. Now we will use the CreateFeature method in order to create a new feature and then populate it with the attributes. The only attribute that we care about now is the geometry or the shape. The shape is actually the polygon we just drew. Write the following code to create the feature:

```
        Dim pWorkspaceEdit As IWorkspaceEdit =
getYharnamWorkspace()
pWorkspaceEdit.StartEditing(True)
pWorkspaceEdit.StartEditOperation()
```

```
        Dim pExcavationFeatureClass As IFeatureClass = getExcavati
onFeatureClass(pWorkspaceEdit)
        Dim pFeature As IFeature = pExcavationFeatureClass.
CreateFeature()
pFeature.Shape = pPolygon
pFeature.Store()

pWorkspaceEdit.StopEditOperation()
pWorkspaceEdit.StopEditing(True)
```

8. Build and run `yharnam.mxd`.

9. Click on the **New Excavation** tool and draw a polygon on the map. Refresh the map. You will see that a new excavation feature is added to the map, as shown in the following screenshot:

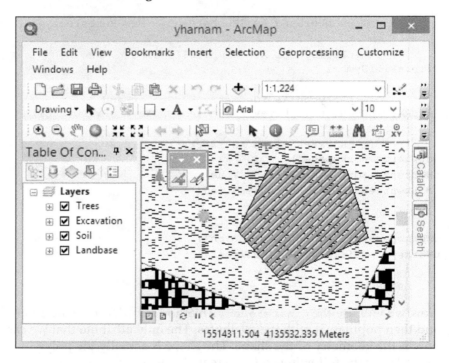

10. Close ArcMap and choose not to save any changes. Reopen `yharnam.mxd` and you will see that the features you created are still there because they are stored in the geodatabase.

11. Close ArcMap and choose not to save any changes.

We have learned how to create features. Now we will learn how to edit excavations as well.

Viewing and editing the excavation information

We have created some excavation features on the map; however, these are merely polygons and we need to extract useful information from them, and display and edit these excavations. For that, we will use the **Excavation Editor** tool to click on an excavation and display the **Excavation Editor** form with the excavation information. Then we will give the ability to edit this information. Follow these steps:

1. If necessary, open Visual Studio Express in administrator mode; we need to do this since our project is actually writing to the registry this time, so it needs administrator permissions. To do that, right-click on **Visual Studio** and click on **Run as administrator**.

2. Go to **File**, then click on **Open Project**, browse to the Yharnam project from C:\ArcGISByExample\yharnam\Code, and click on **Open**.

3. Right-click on frmExcavationEditor.vb and click on **View Code** to view the class code.

4. Add the ArcMapApplication property as shown in the following code so that we can set it from the tool, we will need this at a later stage:

```
Private _application As IApplication
Public Property ArcMapApplication() As IApplication
    Get
        Return _application
    End Get
Set(ByVal value As IApplication)
        _application = value
    End Set
End Property
```

5. Add another method called PopulateExcavation that takes a feature. This method will populate the form fields with the information we get from the excavation feature. We will pass the feature from the **Excavation Editor** tool at a later stage:

```
Public Sub PopulateExcavation(pFeature As IFeature)

End Sub
```

6. According to the following screenshot of the Excavation feature class from ArcCatalog, we can populate three fields from the excavation attributes, these are the design ID, the depth of excavation, and the object ID of the feature:

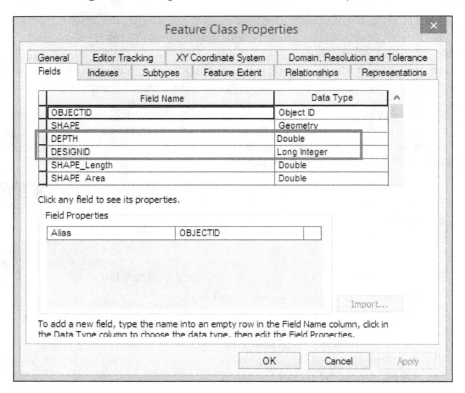

7. Write the following code to populate the design ID, the depth, and the object ID. Note that we used the isDBNull function to check if there is any value stored in those fields. Note that we don't have to do that check for the **OBJECTID** field since it should never be null:

```
Public Sub PopulateExcavation(pFeature As IFeature)

        Dim designID As Long = 0
        Dim dDepth As Double = 0
        If IsDBNull(pFeature.Value(pFeature.Fields.
FindField("DESIGNID"))) = False Then
designID = pFeature.Value(pFeature.Fields.FindField("DESIGNID"))
        End If

        If IsDBNull(pFeature.Value(pFeature.Fields.
FindField("DEPTH"))) = False Then
dDepth = pFeature.Value(pFeature.Fields.FindField("DEPTH"))
```

```
        End If

txtDesignID.Text = designID
txtExcavationDepth.Text = dDepth
txtExcavationOID.Text= pFeature.OID
    End Sub
```

8. What left is the excavation area, which is a bit tricky. To do that, we need to get it from the Shape property of the feature by casting it to the IAreaarcobjects interface and use the area property as follows:

```
    …..
txtExcavationOID.Text = pFeature.OID


    Dim pArea As IArea = pFeature.Shape
txtExcavationArea.Text = pArea.Area
    End Sub
```

9. Now our viewing capability is ready, we need to execute it. Double-click on tlExcavationEditor.vb in order to open the code.

10. We will need the getYharnamWorkspace and getExcavationFeatureClass methods that are in tlNewExcavation.vb. Copy them in tlExcavationEditor.vb.

11. In the onMouseDown event, write the following code to get the feature from the mouse location. This will convert the *x, y* mouse coordinate into a map point and then does a spatial query to find the excavation under this point. After that, we will basically call our excavation editor form and send it the feature to do the work as follows:

```
    Public Overrides Sub OnMouseDown(ByVal Button As Integer,
ByVal Shift As Integer, ByVal X As Integer, ByVal Y As Integer)
        'TODO: Add tlExcavationEditor.OnMouseDown implementation


    Dim pMxdoc As IMxDocument = m_application.Document
    Dim pPoint As IPoint = pMxdoc.ActiveView.ScreenDisplay.
DisplayTransformation.ToMapPoint(X, Y)
    Dim pSFilter As ISpatialFilter = New SpatialFilter
pSFilter.Geometry = pPoint
    Dim pFeatureClass As IFeatureClass = getExcavationFeatureC
lass(getYharnamWorkspace())
    Dim pFCursor As IFeatureCursor = pFeatureClass.
Search(pSFilter, False)
    Dim pFeature As IFeature = pFCursor.NextFeature
    If pFeatureIs Nothing Then Return


    Dim pExcavationEditor As New frmExcavationEditor
```

```
pExcavationEditor.ArcMapApplication = m_application
pExcavationEditor.PopulateExcavation(pFeature)
pExcavationEditor.Show()

    End Sub
```

12. Build and run `yharnam.mxd`.

13. Click on the **Excavation Editor** tool and click on one of the excavations you drew before. You should see that the **Excavation Editor** form pops up with the excavation information; no design ID or depth is currently set, as you can see in the following screenshot:

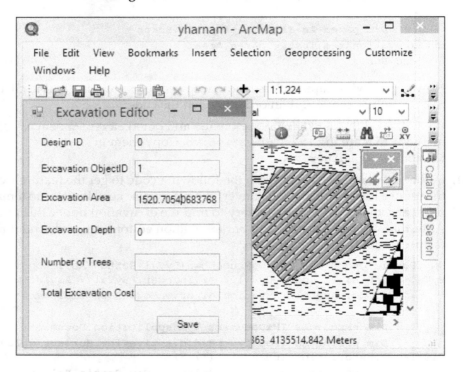

14. Close ArcMap and choose not to save any changes.

15. We will do the final trick to edit the excavation; there is not much to edit here, only the depth. To do that, copy the `getYharnamWorkspace` and `getExcavationFeatureClass` methods that are in `tlNewExcavation.vb`. Copy them in `frmExcavationEditor.vb`. You will get an error in the `m_application.hwnd`, so replace it with `_application.hwnd`, which is the property we set.

16. Right-click on `frmExcavationEditor` and select **View Designer**.

17. Double-click on the **Save** button to generate the `btnSave_Click` method.

18. The user will enter the new depth for the excavation in the `txtExcavationDepth` textbox. We will use this value and store it in the feature. But before that, we need to retrieve that feature using the object ID, start editing, save the feature, and close the session. Write the following code to do so. Note that we have closed the form at the end of the code, so we can open it again to get the new value:

```
Private Sub btnSave_Click(sender As Object, e As EventArgs)
Handles btnSave.Click
        Dim pWorkspaceEdit As IWorkspaceEdit =
getYharnamWorkspace()
        Dim pFeatureClass As IFeatureClass = getExcavationFeatureC
lass(pWorkspaceEdit)

        Dim pFeature As IFeature = pFeatureClass.
GetFeature(txtExcavationOID.Text)

pWorkspaceEdit.StartEditing(True)
pWorkspaceEdit.StartEditOperation()

pFeature.Value(pFeature.Fields.FindField("DEPTH")) =
txtExcavationDepth.Text
pFeature.Store()

pWorkspaceEdit.StopEditOperation()
pWorkspaceEdit.StopEditing(True)
Me.Close
    End Sub
```

19. Build and run `yharnam.mxd`.

20. Click on the **Excavation Editor** tool and click on one of the excavations you drew before. Type a numeric depth value and click on **Save**; this will close the form. Use the **Excavation Editor** tool again to open back the excavation and check if your depth value has been stored successfully.

This is the end of the chapter. You can find the latest code under `B04847_08_Files\yharnam\FinalCode\`.

Summary

In this chapter, you started writing the excavation planning manager, code named Yharnam. In the first part of the chapter, you spent time learning to use the geodatabase editing and preparing the project. You then learned how to use the rubber band tool, which allows you to draw geometries on the map. Using this drawn geometry, you edited the workspace and created a new excavation feature with that geometry. You then learned how to view and edit the excavation feature with attributes.

In the next chapter, you will keep enhancing the Excavation Planning Manager to add more functionality to it. You will work on how to do advanced spatial operations on other feature classes to obtain the cost of an excavation.

Excavation Cost Calculation

9

In the previous chapter, we developed the Excavation Planning Manager, code named Yharnam, we wrote few basic tools to create new excavation and view and edit existing excavation. We learned how to calculate the area of the excavation and read and write attributes to the excavation. However, there were some form fields that we have added such as the number of trees and total cost that we didn't tackle. We will learn how to find the cost based by doing extensive spatial operations in this chapter.

In this chapter, we will discuss the following topics:

- Preparing the excavation cost calculator
- Calculating the soil type removal cost
- Calculating the trees removal cost
- Estimating the final cost

Preparing the excavation cost calculator

Calculating the excavation cost requires us to know of some parameters and basic cost units. These units are prepared by the Yharnam team in the Yharnam geodatabase in the following table:

Item	Item type	Unit	Cost $
ROCK	SOIL	m3	10
SAND	SOIL	m3	5
TYPE1	TREE	Number	5
TYPE2	TREE	Number	10
TYPE3	TREE	Number	50

The cost of removing a single tree depends on its type and is stored in the preceding table. The cost of removing 1 cubic meter of soil depends also on the type of soil. We will develop all these algorithms in the coming sections. Note that the numbers in the table are not real and are here just for the sake of example.

Creating the excavation cost calculator class

We will be writing a lot of code related to the excavation cost calculation, and it is wise to use an object oriented to our advantage. We will create the `ExcavationCostCalculator` class that will hold our method. Follow these steps:

1. If you are continuing from the previous chapter, you can use your final code. If you are starting fresh, copy the entire `yharnam` folder in the supporting files for this chapter `B04847_09_Files\yharnam\` to the `C:\ArcGISByExample\`. The code folder will have the final code from the last chapter.

2. Open Visual Studio Express in administrator mode; we need to do this since our project is actually writing to the registry this time, so it needs administrator permissions. To do that, right-click on **Visual Studio** and click on **Run as administrator**.

3. Go to **File**, then click on **Open Project**, browse to the `Yharnam` project from the `C:\ArcGISByExample\yharnam\Code`, and click on **Open**.

4. Go to the **Project** menu and click on **Add Class**.

5. From **Common Items**, select **Class** and type `ExcavationCostCalculator.vb` in the name.

6. Add the following methods to the class:

```vb
Imports ESRI.ArcGIS.Geodatabase

Public Class ExcavationCostCalculator
    Private _excavationfeature As IFeature

    Public Sub New(pFeature As IFeature)
        _excavationfeature = pFeature
    End Sub

    Public Function TreesCount() As Integer
    End Function

    Public Function TreesRemovalCost() As Double
    End Function
```

```
Public Function SoilRemovalCost() As Double
End Function
End Class
```

7. Note that we have added the excavation feature in the class constructor because we will be needing it. We will also need the getYharnamWorkspace and getExcavationFeatureclass methods, so copy them from tlNewExcavation.vb. You can replace the m_application.hwnd parameter with 0 since we are calling this from a class.

8. Use the same code getExcavationFeatureClass to write the following two functions, getSoilFeatureClass, getTreesFeatureClass, getRemvalCostTable, as follows:

```
Public Function getYharnamWorkspace() As IWorkspace
    Dim pWorkspaceFactory As IWorkspaceFactory = New
FileGDBWorkspaceFactory
    Return pWorkspaceFactory.OpenFromFile("C:\ArcGISByExample\
yharnam\Data\Yharnam.gdb", 0)
End Function


Public Function getExcavationFeatureClass(pWorkspace As
IWorkspace) As IFeatureClass
    Dim pFWorkspace As IFeatureWorkspace = pWorkspace
    Return pFWorkspace.OpenFeatureClass("Excavation")
End Function


    Public Function getSoilFeatureClass(pWorkspace As IWorkspace)
As IFeatureClass
    Dim pFWorkspace As IFeatureWorkspace = pWorkspace
    Return pFWorkspace.OpenFeatureClass("Soil")
End Function


    Public Function getTreesFeatureClass(pWorkspace As IWorkspace)
As IFeatureClass
    Dim pFWorkspace As IFeatureWorkspace = pWorkspace
    Return pFWorkspace.OpenFeatureClass("Trees")
End Function


    Public Function getRemovalCostTable(pWorkspace As IWorkspace)
As ITable
    Dim pFWorkspace As IFeatureWorkspace = pWorkspace
    Return pFWorkspace.OpenTable("RemovalCost")
End Function
```

9. Import any missing libraries and then save your solution.

Calculating the soil type removal cost

To calculate the cost of removing a particular soil type, we need to first find the underlying soil feature intersecting the excavation. Then we find the intersecting area between the excavation feature and the soil feature, calculate the area, and multiply it by the depth of the excavation. This will give us the total volume of the excavation. We then query the removal cost table to find the cost of removing the 1 m3 of that particular soil and then multiply it by the excavation volume. Note that an excavation might pass between two soil type areas, which will result in multiple intersections from different soil types. Take a look at the following example drawing:

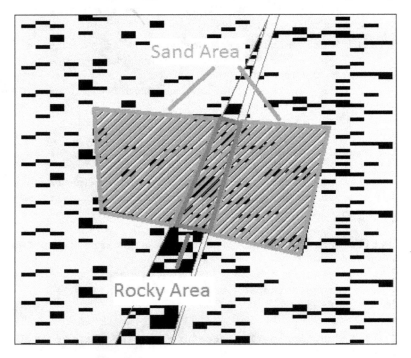

Follow these steps to calculate the cost of removing the soil:

1. If necessary, open the `Yharnam` project as administrator.
2. Edit the `ExcavationCostCalculator` class.

3. First we will need to find all of the soil features that intersect with the excavation. We will do a spatial query for that. Write the following code in the `SoilRemovalCost`:

```
Public Function SoilRemovalCost() As Double
        Dim pWorkspace As IWorkspace = getYharnamWorkspace()
        Dim dTotalSoilRemovalCostas Double =0
        Dim pSoilFC As IFeatureClass =
getSoilFeatureClass(pWorkspace)
        Dim pExcavationFC As IFeatureClass = getExcavationFeatureC
lass(pWorkspace)
        Dim pSFilter As ISpatialFilter = New SpatialFilter
pSFilter.SpatialRel = esriSpatialRelEnum.esriSpatialRelIntersects
pSFilter.Geometry = _excavationfeature.Shape

        Dim pFCursor As IFeatureCursor = pSoilFC.Search(pSFilter,
False)
        Dim pSoilFeature As IFeature = pFCursor.NextFeature

        Do UntilpSoilFeature Is Nothing

pSoilFeature = pFCursor.NextFeature
        Loop
    End Function
```

4. For each soil feature, we need find the intersection area geometry. For that, we use the topological operator:

```
Do UntilpSoilFeature Is Nothing

        Dim pTopologicalOperator As ITopologicalOperator = _
excavationfeature.Shape
        Dim pIntersection As IGeometry = pTopologicalOperator.
Intersect(pSoilFeature.Shape, esriGeometryDimension.
esriGeometry2Dimension)

pSoilFeature = pFCursor.NextFeature
        Loop
```

5. Now that we have a soil intersection, we need to calculate the area and multiply it by the depth to find the volume of excavation for that particular cost:

```
Do UntilpSoilFeature Is Nothing

 Dim pTopologicalOperator As ITopologicalOperator = _
excavationfeature.Shape
Dim pIntersection As IGeometry = pTopologicalOperator.
Intersect(pSoilFeature.Shape, esriGeometryDimension.
esriGeometry2Dimension)
Dim pArea As IArea = pIntersection
Dim dExcavationDepth As Double
        If IsDBNull(_excavationfeature.Value(_
excavationfeature.Fields.FindField("DEPTH"))) Then
dExcavationDepth = 0
        Else
dExcavationDepth = _excavationfeature.Value(_excavationfeature.
Fields.FindField("DEPTH"))
        End If

Dim dSoilExcavationVolume As Double = Math.Abs(pArea.Area) *
dExcavationDepth


pSoilFeature = pFCursor.NextFeature
Loop
```

6. Now that we have the volume, we need to hit the cost removal table and get the cost per unit and multiply it by the volume. Then we simply add this to the total cost. We will apply the same algorithm for each soil feature. Note that we have to finally return the value of total excavation cost:

```
Do UntilpSoilFeature Is Nothing

        Dim pTopologicalOperator As ITopologicalOperator = _
excavationfeature.Shape
        Dim pIntersection As IGeometry = pTopologicalOperator.
Intersect(pSoilFeature.Shape, esriGeometryDimension.
esriGeometry2Dimension)
        Dim pArea As IArea = pIntersection

Dim dExcavationDepth As Double
        If IsDBNull(_excavationfeature.Value(_
excavationfeature.Fields.FindField("DEPTH"))) Then
dExcavationDepth = 0
        Else
```

```
dExcavationDepth = _excavationfeature.Value(_excavationfeature.
Fields.FindField("DEPTH"))
          End If


          Dim dSoilExcavationVolume As Double = Math.Abs(pArea.
Area) * dExcavationDepth

          Dim sSoilType As String = pSoilFeature.
Value(pSoilFeature.Fields.FindField("SOILTYPE"))
Dim pRemovalCost As ITable = getRemovalCostTable(pWorkspace)
          Dim pQfilter As IQueryFilter = New QueryFilter
pQfilter.WhereClause = "ITEM = '" &sSoilType& "'"
          Dim pCursor As ICursor = pRemovalCost.Search(pQfilter,
False)
          Dim pRow As IRow = pCursor.NextRow


          Dim dCost As Double = pRow.Value(pRow.Fields.
FindField("COST"))

dTotalSoilRemovalCost = dTotalSoilRemovalCost + dCost *
dSoilExcavationVolume

pSoilFeature = pFCursor.NextFeature
      Loop
Return dTotalSoilRemovalCost
```

7. Now we need to call our new method from the `frmExcavationEditor` form. Add the following lines at the end of the `PopulateExcavation` method to set the total cost of excavation:

```
Public Sub PopulateExcavation(pFeature As IFeature)

      Dim designID As Long = 0
      Dim dDepth As Double = 0
      If IsDBNull(pFeature.Value(pFeature.Fields.
FindField("DESIGNID"))) = False Then
designID = pFeature.Value(pFeature.Fields.FindField("DESIGNID"))
      End If


      If IsDBNull(pFeature.Value(pFeature.Fields.
FindField("DEPTH"))) = False Then
dDepth = pFeature.Value(pFeature.Fields.FindField("DEPTH"))
      End If

txtDesignID.Text = designID
txtExcavationDepth.Text = dDepth
```

```
txtExcavationOID.Text = pFeature.OID

        Dim pArea As IArea = pFeature.Shape
txtExcavationArea.Text = pArea.Area


Dim pExcavationCostCalculator As New ExcavationCostCalculator(pFea
ture)
txtTotalCost.Text = pExcavationCostCalculator.SoilRemovalCost()

    End Sub
```

8. Build your solution and run `Yharnam.mxd`.

9. Activate the Excavation Editor and select an excavation; if you don't have any, create one. For instance, this particular excavation spans three different soil areas of 228 meter square with a depth of 3 meters. The total cost is $3719, as illustrated in the following screenshot:

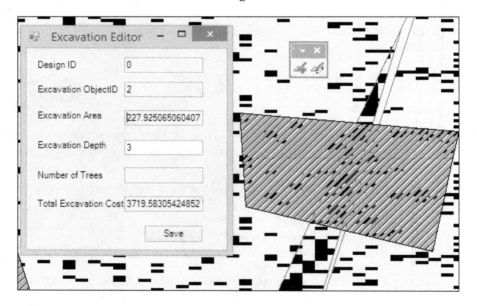

10. Close ArcMap and choose not to save any changes.

Calculating the tree removal cost

In this section, we will continue working on our cost calculator class and add more functionalities to it. We will add the ability to calculate the number of trees under a certain excavation and then we will also calculate the cost of removing those trees based on their type.

Calculating the number of trees

To calculate the number of trees, we will need to perform a spatial query on the tree feature class to find all trees under the current selected excavation.

1. If necessary, open Visual Studio Express in administrator mode; we need to do this since our project is actually writing to the registry this time, so it needs administrator permissions. To do that, right-click on **Visual Studio** and click on **Run as administrator**.

2. Go to **File**, then click on **Open Project**, browse to the Yharnam project from the C:\ArcGISByExample\yharnam\Code, and click on **Open**.

3. Double-click on ExcavationCostCalculator.vb to edit the code.

4. We will first need to get the tree feature class—we have a function that returns the tree feature class for us—we then need to create a spatial filter on the tree feature class using the excavation geometry as input. We will then use the FeatureCount method and pass the spatial filter to it to return the number of trees. Write the following in the TreesCount method:

```
Public Function TreesCount() As Integer
    Dim pWorkspace As IWorkspace = getYharnamWorkspace()
    Dim pTreeFeatureClass As IFeatureClass = getTreesFeatureCl
ass(pWorkspace)
    Dim pSFilter As ISpatialFilter = New SpatialFilter
pSFilter.Geometry = _excavationfeature.Shape
pSFilter.SpatialRel = esriSpatialRelEnum.esriSpatialRelIntersects

    Return pTreeFeatureClass.FeatureCount(pSFilter)
End Function
```

5. Go to frmExcavationEditor and call the getTreeCount method on your PopulateExcavation method as follows:

```
Dim pExcavationCostCalculator As New ExcavationCostCalcula
tor(pFeature)
txtTotalCost.Text = pExcavationCostCalculator.SoilRemovalCost()
txtTreeCount.Text = pExcavationCostCalculator.TreesCount()

End Sub
```

6. Build your solution; if it fails, make sure you have run the solution as administrator.

7. Run `yharnam.mxd`.

8. Activate the Excavation Editor and select an excavation. The tree count represents the green factor; the more it is, the more careful one should be when removing a green area, as shown in the following screenshot:

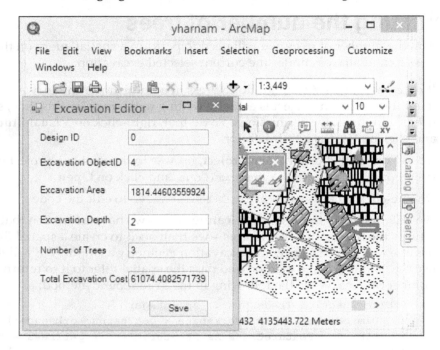

9. Close ArcMap and choose not to save the changes.

Calculating the tree removal cost

To calculate the cost of removing each tree, we should know how much each tree costs to remove based on its type. Follow these steps to calculate the cost of removing the trees:

1. If necessary, open Visual Studio Express in administrator mode; we need to do this since our project is actually writing to the registry this time, so it needs administrator permissions. To do that, right-click on **Visual Studio** and click on **Run as administrator**.

2. Go to **File**, then click on **Open Project**, browse to the `Yharnam` project from the `C:\ArcGISByExample\yharnam\Code`, and click on **Open**.

3. Double-click on `ExcavationCostCalculator.vb` to edit the code.

4. In `TreesRemovalCost`, we will write the logic to calculate the cost of removing each tree. This is easier than the soil type calculation since we don't need to find an intersection or work with areas. We just need to get the cost of the tree removal based on its type by querying the removal cost table. Write the following code to get the cursor of all trees under the excavation:

```
Public Function TreesRemovalCost() As Double
    Dim pWorkspace As IWorkspace = getYharnamWorkspace()
    Dim dTotalTreeRemovalCost As Double = 0
    Dim pTreesFC As IFeatureClass = getTreesFeatureClass(pWork
space)

    Dim pSFilter As ISpatialFilter = New SpatialFilter
pSFilter.Geometry = _excavationfeature.Shape
pSFilter.SpatialRel = esriSpatialRelEnum.esriSpatialRelIntersects
    Dim pFCursor As IFeatureCursor = pTreesFC.Search(pSFilter,
False)
    Dim pTreeFeature As IFeature = pFCursor.NextFeature

    Do UntilpTreeFeature Is Nothing
        Dim sTreeType As String = pTreeFeature.
Value(pTreeFeature.Fields.FindField("TREETYPE"))
        Dim dTreeCost As Double = 0


dTotalTreeRemovalCost = dTotalTreeRemovalCost + dTreeCost

pTreeFeature = pFCursor.NextFeature
        Loop

        Return dTotalTreeRemovalCost
    End Function
```

5. Now we need to get the tree type and query the removal cost table to get the cost of removing that tree as follows:

```
Public Function TreesRemovalCost() As Double
    Dim pWorkspace As IWorkspace = getYharnamWorkspace()
    Dim dTotalTreeRemovalCost As Double = 0
    Dim pTreesFC As IFeatureClass = getTreesFeatureClass(pWork
space)

    Dim pSFilter As ISpatialFilter = New SpatialFilter
pSFilter.Geometry = _excavationfeature.Shape
```

```
        pSFilter.SpatialRel = esriSpatialRelEnum.esriSpatialRelIntersects
            Dim pFCursor As IFeatureCursor = pTreesFC.Search(pSFilter,
False)
            Dim pTreeFeature As IFeature = pFCursor.NextFeature

            Do UntilpTreeFeature Is Nothing
                Dim sTreeType As String = pTreeFeature.
Value(pTreeFeature.Fields.FindField("TREETYPE"))
                Dim dTreeCost As Double = 0

    Dim pRemovalCost As ITable = getRemovalCostTable(pWorkspace)
                Dim pQfilter As IQueryFilter = New QueryFilter
    pQfilter.WhereClause = "ITEM = '" &sTreeType& "'"
                Dim pCursor As ICursor = pRemovalCost.Search(pQfilter,
False)
                Dim pRow As IRow = pCursor.NextRow

    dTreeCost = pRow.Value(pRow.Fields.FindField("COST"))

    dTotalTreeRemovalCost = dTotalTreeRemovalCost + dTreeCost

    pTreeFeature = pFCursor.NextFeature
            Loop

            Return dTotalTreeRemovalCost
        End Function
```

6. Now we need to call the tree removal cost from the `frmExcavationEditor`
   and add it to the cost of removing the soil as follows:

```
    Public Sub PopulateExcavation(pFeature As IFeature)

            Dim designID As Long = 0
            Dim dDepth As Double = 0
            If IsDBNull(pFeature.Value(pFeature.Fields.
FindField("DESIGNID"))) = False Then
    designID = pFeature.Value(pFeature.Fields.FindField("DESIGNID"))
            End If

            If IsDBNull(pFeature.Value(pFeature.Fields.
FindField("DEPTH"))) = False Then
    dDepth = pFeature.Value(pFeature.Fields.FindField("DEPTH"))
            End If
```

```
txtDesignID.Text = designID
txtExcavationDepth.Text = dDepth
txtExcavationOID.Text = pFeature.OID


        Dim pArea As IArea = pFeature.Shape
txtExcavationArea.Text = Math.Abs(pArea.Area)



        Dim pExcavationCostCalculator As New ExcavationCostCalcula
tor(pFeature)
txtTotalCost.Text = pExcavationCostCalculator.SoilRemovalCost() +
pExcavationCostCalculator.TreesRemovalCost()
txtTreeCount.Text = pExcavationCostCalculator.TreesCount()


        End Sub
```

7. Build and run `yharnam.mxd`.

8. Activate the Excavation Editor and select an excavation. Note that the cost has now increased. In the following screenshot, we have two Type 2 trees and one Type 1. According to the removal cost table, this will be *2*10 + 5*, which is an addition of $25 dollars only, from $61,074 to $61,099, as illustrated in the following screenshot:

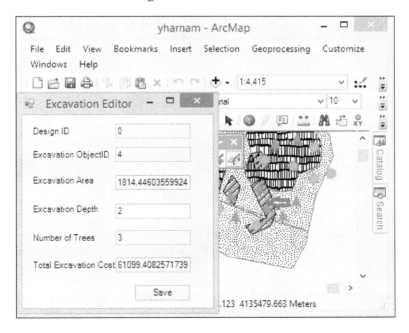

9. Close ArcMap and choose not to save any changes.

Estimating the final cost

We have completed our calculation model, and we just need to add some finishing touches. First we will add a new method to `ExcavationCostcalculator` called `TotalCost`, which will simply return the sum of cost of removing the trees plus the cost of removing the soil. We will also add the cost breakdown in a small message box. Follow these steps:

1. If necessary, open Visual Studio Express in administrator mode; we need to do this since our project is actually writing to the registry this time, so it needs administrator permissions. To do that, right-click on **Visual Studio** and click on **Run as administrator**.

2. Go to **File**, then click on **Open Project**, browse to the `Yharnam` project from the `C:\ArcGISByExample\yharnam\Code`, and click on **Open**.

3. Double-click on the `ExcavationCostCaculator.vb` class to edit it and add the following method:

```
Public Function TotalExcavationCost() As Double
    Return SoilRemovalCost() + TreesRemovalCost()
End Function
```

4. Right-click on `frmExcavationEditor.vb` and click on `View Code` to view the class code.

5. Set `txtTotalCost.text` to be the new method you just wrote. We can also fix the round the area and the cost to the nearest decimal point using the `Math.Round` method. We also add the units as well:

```
Public Sub PopulateExcavation(pFeature As IFeature)

        Dim designID As Long = 0
        Dim dDepth As Double = 0
        If IsDBNull(pFeature.Value(pFeature.Fields.
FindField("DESIGNID"))) = False Then
designID = pFeature.Value(pFeature.Fields.FindField("DESIGNID"))
        End If

        If IsDBNull(pFeature.Value(pFeature.Fields.
FindField("DEPTH"))) = False Then
dDepth = pFeature.Value(pFeature.Fields.FindField("DEPTH"))
        End If

txtDesignID.Text = designID
txtExcavationDepth.Text = dDepth
txtExcavationOID.Text = pFeature.OID
```

```
        Dim pArea As IArea = pFeature.Shape
txtExcavationArea.Text = Math.Round(Math.Abs(pArea.Area), 1)& "
m2"



        Dim pExcavationCostCalculator As New ExcavationCostCalcula
tor(pFeature)
txtTotalCost.Text=  "$" &Math.Round(pExcavationCostCalculator.
TotalExcavationCost, 1)
txtTreeCount.Text = pExcavationCostCalculator.TreesCount()

    End Sub
```

6. Build and run `yharnam.mxd`.

7. Click on the Excavation Editor tool and click on one of the excavations you drew before to make sure everything is in order, as shown in the following screenshot:

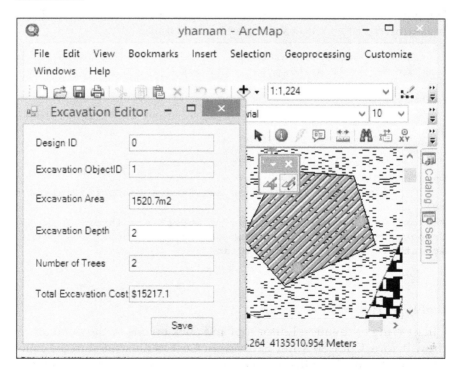

8. Close ArcMap and choose not to save any changes.

9. Right-click on the `frmExcavationEditor` and select **View Designer**.

10. Double-click on the `lblTotalCost` label to generate the `lblTotalCost_Click` method.

11. When the user clicks on this label, we will simply breakdown the cost into two parts: the cost of removing the trees plus the cost of removing the soil. For that, we need to retrieve the excavation feature since we don't have it here. We will use the object ID to do so:

```
Private Sub lblTotalCost_Click(sender As Object, e As
EventArgs) Handles lblTotalCost.Click
        Dim pWorkspaceEdit As IWorkspaceEdit =
getYharnamWorkspace()
        Dim pFeatureClass As IFeatureClass = getExcavationFeatureC
lass(pWorkspaceEdit)


        Dim pFeature As IFeature = pFeatureClass.
GetFeature(txtExcavationOID.Text)


    End Sub
```

We will use our calculator and pass the excavation feature to retrieve the breakdown cost as follows and display a message:

```
Private Sub lblTotalCost_Click(sender As Object, e As
EventArgs) Handles lblTotalCost.Click
        Dim pWorkspaceEdit As IWorkspaceEdit =
getYharnamWorkspace()
        Dim pFeatureClass As IFeatureClass = getExcavationFeatureC
lass(pWorkspaceEdit)


        Dim pFeature As IFeature = pFeatureClass.
GetFeature(txtExcavationOID.Text)
        Dim pCostCalculator As New ExcavationCostCalculator(pFeatu
re)
MsgBox("Cost of Removing Trees: " &pCostCalculator.
TreesRemovalCost&vbCrLf& "Cost of Removing Soil: "
&pCostCalculator.SoilRemovalCost)


    End Sub
```

12. Build and run `yharnam.mxd`.

13. Click on the Excavation Editor tool and click on one of the excavations you drew before. Click on **Total Excavation Cost** as illustrated in the following screenshot. Note how the total excavation cost is in fact equal to the tree removal cost plus the excavation cost.

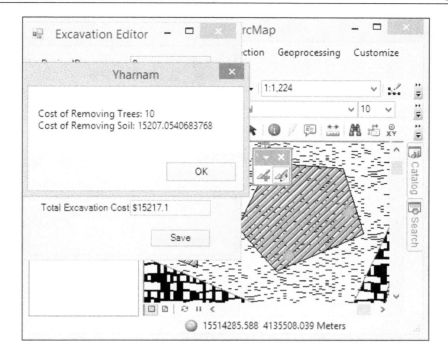

Summary

In this chapter, you started to add more rich functionality using the Excavation Planning Manager. In the first part of the chapter, you spent time preparing the cost calculator class, which helped in centralizing the cost calculation code. Then you learned how to calculate the cost of removing a soil by calculating the intersecting area between the excavation and the soil feature class. You also learned how to find the cost of removing the trees that is underlying an excavation and how the number of trees affect the green factor of the area you are excavating.

In the next chapter, you will put the final touches on Yharnam to create excavation designs. Each design will have multiple excavations and the design will have a final cost. The engineer can create multiple designs and compare the cost to make the final decision.

10
Saving and Retrieving Excavation Designs

This is the last, and the longest, chapter in this book. It is here where you will feel that you are working with a real project. You almost snap from the feeling of example development into the realistic project requirements. In this chapter, we will introduce the concept of an excavation design. A construction designer can create a new design and add multiple excavations that belong to one design. The Excavation Manager should then calculate the total cost of all excavations. This way the designer can compare multiple designs and their cost, and decide which one should be implemented. There will be reporting, editing, and deleting involved, so stay tuned and get ready for some serious ArcGIS coding.

In this chapter, we will discuss the following topics:

- Preparing the design table
- Creating the design manager button
- Opening existing designs
- Deleting designs
- Generating the excavation design report
- Searching for designs

Preparing the design table

First of all, we will need to create the design table in which all designs will be stored. This table currently does not exist in the Yharnam geodatabase. The table should include `DesignID` and `DesignDate`. To create the table, follow these steps:

1. Copy the entire `yharnam` folder in the supporting files for this chapter `B04847_10_Files\yharnam` to `C:\ArcGISByExample\yharnam`. You don't have to do this step if you completely finished *Chapter 9, Excavation Cost Calculation*; this will simply take the latest change that has been made in that chapter.

2. Open ArcCatalog application and browse to the `yharnam` geodatabase in `C:\ArcGISByExample\yharnam\Data\Yharnam.gdb`.

3. Right-click on the `Yharnam.gdb` geodatabase, go to **New**, and then click on **Table**, as illustrated in the following screenshot:

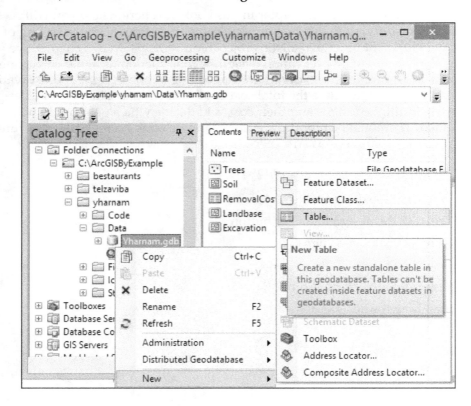

4. In the **New Table** dialog, type `Design` in both **Name** and **Alias** and click on **Next**.

5. Select default configuration keyword and click on **Next**.

6. Add two fields, DESIGNID of type Long Integer and DESIGNDATE of type Date, as shown in the following screenshot:

7. Click on **Finish** to create the table.

Note that we didn't store the design cost in this table because design cost keep changing based on the design parameters, such as adding new excavations and new trees, so it should always calculate the new cost when you open the design instead of storing an old value of the cost.

Creating the design manager

The design manager will help us manage the excavation designs. We create this form to add new designs, edit existing designs, and open (and even delete) designs. This will be our main focus point throughout this chapter.

Adding the design manager button

First we will need to add an ArcMap button to our Excavation Manager Toolbar to show the design manager. Follow these steps to create the button:

1. Open Visual Studio Express in administrator mode; we need to do this since our project is actually writing to the registry this time, so it needs administrator permissions. To do that, right-click on **Visual Studio** and click on **Run as administrator**.

2. Go to **File**, then click on **Open Project**, browse to the Yharnam project from the C:\ArcGISByExample\yharnam\Code, and click on **Open**.

3. Add an `ArcMapBase` command button and name it `cmDesignManager.vb`. Set `m_category` to `Yharnam`, `m_name` to `Yharnam_DesignManager`, and `m_caption`, `m_message`, and `m_tooltip` to `Design manager`. By now you should know how to do this.

4. Double-click and edit the `cmDesignManager.bmp` picture and set it to the icon located in `C:\ArcGISByExample\Icons\excavation_manager.png`.

5. Add the button to the `tbYharnam` toolbar by using its Progid.

6. Save your project.

Preparing the design manager form

Here is where we prepare the design manager form and have all different fields and controls that we will be using during this chapter. Follow these steps to add the design manager form:

1. If necessary, open the `Yharnam` project as administrator.

2. From the project menu, go to **Add Windows Form** and name it `frmDesignManager`.

3. Add the following controls with their attributes to your form design view:

Name	Type	Properties
`lblDesignID`	Label	Text: Design ID
`lblDesignDate`	Label	Text: Design Date
`lblExcavations`	Label	Text: Excavations
`lblTrees`	Label	Text: Trees
`lblTotalArea`	Label	Text: Total Area
`lblTreeCount`	Label	Text: Number of Trees
`lblExcavationCost`	Label	Text: Excavation Cost
`lblTreeRemovalCost`	Label	Text: Trees Removal Cost
`lblTotalCost`	Label	Text: Total Cost Font:Bold
`txtDesignID`	Text	Readonly
`txtDesignDate`	Text	Readonly
`txtTotalArea`	Text	Readonly
`txtTreeCount`	Text	Readonly
`txtExcavationCost`	Text	Readonly
`txtTreeRemovalCost`	Text	Readonly
`txtTotalcost`	Text	Readonly

Name	Type	Properties
lstExcavations	ListBox	
lstTrees	ListBox	
btnAddExcavation	Button	Text: Add Excavation
btnDeleteExcavation	Button	Text: Delete Excavation
btnNewDesign	Button	Text: New Design
btnOpenDesign	Button	Text: Open Design
btnDeleteDesign	Button	Text: Delete Design
btnReport	Button	Text: Report

4. Your form should look as follows when you have added all the controls:

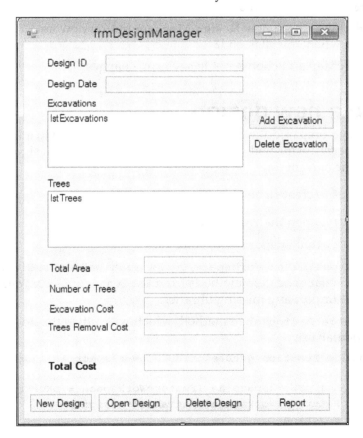

5. View the code of frmDesignManager and add ArcMapApplication of type IApplication as we have learned in previous chapters.

6. Now that we have prepared the form, we need to call it when the user clicks on the **Design Manager** command. We will also need to set `ArcMapApplication` since we will need it. Add this code on the `onClick` method of `cmDesignerManager.vb`:

```
Public Overrides Sub OnClick()
    Dim pDesignManager As New frmDesignManager
    pDesignManager.ArcMapApplication = m_application
    pDesignManager.Show()


End Sub
```

7. Build your `yahrnam` solution and run `yharnam.mxd` in `C:\ArcGISByExample\Data\yharnam.mxd`.

8. Note that we have a new **Design Manager** button now and it displays the form when you click on it.

9. Close ArcMap and choose not to save any changes.

Creating a new design

Now that we have prepared the design manager form, it is time to add some code to create our first design. This will write a new record with a new design ID and set the design date as the current date.

Follow these steps to create a new design:

1. If necessary, open the `Yharnam` project as administrator.

2. Edit `frmDesignManager.vb` in code view.

3. Copy the `getYharnamWorkspace`, `getExcavationFeatureClass`, and `getTreesFeatureClass` methods from `ExcavationCostCalculator` to the form and import any missing libraries.

4. Write the `getDesignTable` method, which will return the design table from the geodatabase:

```
Public Function getDesignTable(pWorkspace As IWorkspace) As
ITable
    Dim pFWorkspace As IFeatureWorkspace = pWorkspace
    Return pFWorkspace.OpenTable("Design")
End Function
```

5. Open the `frmDesignManager` design view and double-click on the **New Design** button to generate the `btnNewDesign_Click` method.

6. Here we will need to get hold of the design table and start editing. Create a new row as follows:

```
    Private Sub btnNewDesign_Click(sender As Object, e As EventArgs)
Handles btnNewDesign.Click
        Dim pWorkspaceEdit As IWorkspaceEdit =
getYharnamWorkspace()
        Dim pDesignTable As ITable =
getDesignTable(pWorkspaceEdit)

        pWorkspaceEdit.StartEditing(True)
        pWorkspaceEdit.StartEditOperation()

        Dim pNewDesignRow As IRow = pDesignTable.CreateRow()

        pWorkspaceEdit.StopEditOperation()
        pWorkspaceEdit.StopEditing(True)
    End Sub
```

7. We will now use the object ID which is autogenerated as the design ID. We will also save the current date into the DesignDate field and save our design. Then we'll update the text fields in the form, as follows:

```
    Private Sub btnNewDesign_Click(sender As Object, e As EventArgs)
Handles btnNewDesign.Click
        Dim pWorkspaceEdit As IWorkspaceEdit =
getYharnamWorkspace()
        Dim pDesignTable As ITable =
getDesignTable(pWorkspaceEdit)

        pWorkspaceEdit.StartEditing(True)
        pWorkspaceEdit.StartEditOperation()

        Dim pNewDesignRow As IRow = pDesignTable.CreateRow()
pNewDesignRow.Value(pNewDesignRow.Fields.FindField("DESIGNID")) =
pNewDesignRow.OID
        pNewDesignRow.Value(pNewDesignRow.Fields.
FindField("DESIGNDATE")) = Now
        pNewDesignRow.Store()

        pWorkspaceEdit.StopEditOperation()
        pWorkspaceEdit.StopEditing(True)

        txtDesignDate.Text = Now
        txtDesignID.Text = pNewDesignRow.OID
    End Sub
```

8. Build your yahrnam solution and run `yharnam.mxd`.

9. Click on the **Design Manager** button to show the design manager.

10. In the **Design Manager** form, click on the **New Design** button and note how you will get a new design ID and date in the field, as shown in the following screenshot:

11. Optionally, verify from ArcCatalog that a new record has been created in the design table.

12. Close ArcMap and choose not to save any changes.

Saving multiple excavations

Now that we have created a new design, it is time to edit that design and assign excavations to that design. To do that, we will need to reuse our `tlNewExcavation.vb` class, which basically creates a new excavation. However, we need to change the logic so that excavation also saves the current design ID. Follow these steps to start working:

1. If necessary, open the `Yharnam` project as administrator.

2. Edit the `tlNewExcavation.vb` class and add the following variable as `Shared`. Shared variables are accessible during the lifetime of a class instead of an object. That means, no matter how many objects we create of this class, they will all point to the same variable value. We set the value to `-1`, which means no designs are currently open:

```
Public NotInheritable Class tlNewExcavation
    Inherits BaseTool

 Public Shared DesignID As Long=-1
 ...
 ...
```

3. In the `onMouseDown` method, set the excavation feature `DESIGNID` field to the `DesignID` shared variable as follows:

```
Public Overrides Sub OnMouseDown(ByVal Button As Integer, ByVal
Shift As Integer, ByVal X As Integer, ByVal Y As Integer)

        Dim pDocument As IMxDocument = m_application.Document

        Dim pRubberBand As IRubberBand = New RubberPolygonClass

        Dim pFillSymbol As ISimpleFillSymbol = New
SimpleFillSymbol

        Dim pPolygon As IGeometry = pRubberBand.
TrackNew(pDocument.ActiveView.ScreenDisplay, pFillSymbol)

        Dim pWorkspaceEdit As IWorkspaceEdit =
getYharnamWorkspace()
        pWorkspaceEdit.StartEditing(True)
        pWorkspaceEdit.StartEditOperation()

        Dim pExcavationFeatureClass As IFeatureClass = getExcavati
onFeatureClass(pWorkspaceEdit)
```

```
        Dim pFeature As IFeature = pExcavationFeatureClass.
CreateFeature()
        pFeature.Shape = pPolygon
        pFeature.Value(pFeature.Fields.FindField("DESIGNID")) =
DesignID
        pFeature.Store()

        pWorkspaceEdit.StopEditOperation()
        pWorkspaceEdit.StopEditing(True)

    End Sub
```

4. This will make sure that each excavation drawn has a design ID. Now we need to go back to design manager, set the design ID of the new excavation tool, and then activate the tool. In the `frmDesignManager` design mode, double-click on the **Add Excavation** button to generate the `btnAddExcavation_Click` method.

5. Set the design ID from the `txtDesignID` field to the `tlNewExcavation` tool as follows:

```
 Private Sub btnAddExcavation_Click(sender As Object, e As
EventArgs) Handles btnAddExcavation.Click
        tlNewExcavation.DesignID = txtDesignID.Text
    End Sub
```

6. Now we will need to activate the tool. Write the following code to activate our **New Excavation** tool using its prog ID:

```
Private Sub btnAddExcavation_Click(sender As Object, e As
EventArgs) Handles btnAddExcavation.Click
        tlNewExcavation.DesignID = txtDesignID.Text

        Dim pUID As New UID
        Dim pCmdItem As ICommandItem
        pUID.Value = "Yharnam.tlNewExcavation"
        pUID.SubType = 3
        pCmdItem = _application.Document.CommandBars.Find(pUID)

        _application.CurrentTool = pCmdItem

    End Sub
```

7. The new excavation tool will remain active until we change the tool manually, so we must reset the tool to the original cursor when we close the form. Add the following event on `FormClosed` to set the current tool to `Nothing`:

```
Private Sub frmDesignManager_FormClosed(sender As Object, e As
Windows.Forms.FormClosedEventArgs) Handles Me.FormClosed
        _application.CurrentTool = Nothing
    End Sub
```

8. Build your solution and run `yharnam.mxd`.

9. Open the Design Manager.

10. From the **Design Manager** form, click on **New Design**. You should get a new design ID, mine is currently `3`.

11. Click on **Add Excavation**; this will minimize your form and will allow you to draw on the map-- draw an excavation. Refresh the map to show your new excavation.

12. Close the **Design Manager** form.

13. Activate the Excavation Editor and click on one of your excavations. You should now see the design ID. Set a depth of `1` and click on **Save**.

14. Edit the excavation again to see the calculated cost, as illustrated in the following screenshot:

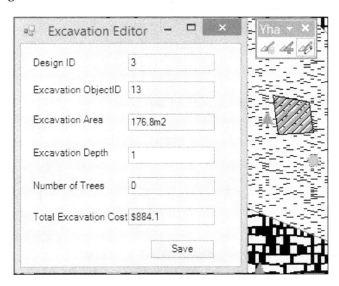

15. Now we will create a new design and draw more than one excavation. Open the Design Manager and click on **New Design**.

16. Click on **Add Excavation** and draw four excavations.

17. Using the Excavation Editor, verify that all four excavations have the same design ID.

18. Close ArcMap and choose not to save any changes.

Opening existing designs

In this section, we will learn how to open an existing design by specifying the design ID. We will write a method called `LoadDesign` which will query the `Design` table, verify that the design exists, retrieve the date, populate the form with all the details, and filter the map to show only excavation for that particular design.

The LoadDesign method

We will create the `LoadDesign` method that will be responsible for querying and loading all design details. Follow these steps to add the method:

1. If necessary, open Visual Studio Express in administrator mode; we need to do this since our project is actually writing to the registry this time, so it needs administrator permissions. To do that, right-click on **Visual Studio** and click on **Run as administrator**.

2. Go to **File**, then click on **Open Project**, browse to the `Yharnam` project from `C:\ArcGISByExample\yharnam\Code`, and click on **Open**.

3. Right-click on `frmDesignManager.vb` and select **View Code**.

4. Write the `LoadDesign` method, which accepts `designid` as a parameter, as follows:

```
Public Sub LoadDesign(designid As Long)

    End Sub
```

5. We will now write the code to query the `Design` table and retrieve the record matching the design ID in the parameter. If no records are returned, we will display the message, "Design was not found". If a record is found, we will populate the design ID and design date fields, as illustrated in the following code:

```
Public Sub LoadDesign(designid As Long)
        Dim pWorkspace As IWorkspace = getYharnamWorkspace()
        Dim pTable As ITable = getDesignTable(pWorkspace)
        Dim pQFilter As IQueryFilter = New QueryFilter
        pQFilter.WhereClause = "DESIGNID = " & designid
```

```
      Dim pCursor As ICursor = pTable.Search(pQFilter, False)
      Dim pRow As IRow = pCursor.NextRow

      If pRow Is Nothing Then
          MsgBox("Design was not found!")
          Exit Sub
      End If



      txtDesignID.Text = pRow.Value(pRow.Fields.
  FindField("DESIGNID"))
      txtDesignDate.Text = pRow.Value(pRow.Fields.
  FindField("DESIGNDATE"))



  End Sub
```

Loading excavations

We will continue with writing more code in our LoadDesign method. This is where
we populate the excavations into the excavations list:

1. If necessary, open Visual Studio Express in administrator mode; we need
   to do this since our project is actually writing to the registry this time, so it
   needs administrator permissions. To do that, right-click on **Visual Studio**
   and click on **Run as administrator**.

2. Go to **File**, then click on **Open Project**, browse to the Yharnam project from
   the C:\ArcGISByExample\yharnam\Code, and click on **Open**.

3. We will need to query the Excavation feature class and retrieve all features
   having this designID. We will then load these excavations into the list. The
   code is shown as follows; note that we can use the same query filter as the
   table:

```
      Public Sub LoadDesign(designid As Long)
  .. . .
  . . . .
      Dim pFeatureClass As IFeatureClass = getExcavationFeatureC
  lass(pWorkspace)
      Dim pFCursor As IFeatureCursor = pFeatureClass.
  Search(pQFilter, False)
      Dim pFeature As IFeature = pFCursor.NextFeature
      lstExcavations.Items.Clear()
```

```
        Do Until pFeature Is Nothing
            lstExcavations.Items.Add(pFeature.OID)
            pFeature = pFCursor.NextFeature
        Loop
End Sub
```

4. We can now test our `LoadDesign` method, but before we do that, we need to actually call the method, so open **Form Designer** in `frmDesignManager.vb`.

5. Double-click on the **Open Design** button to generate the `btnOpenDesign_Click`.

6. We need to ask the user to input `designID`. We will use the simple `InputBox` dialog, which prompts the user for an entry as follows:

```
    Private Sub btnOpenDesign_Click(sender As Object, e As
EventArgs) Handles btnOpenDesign.Click
        Dim designid As Long = InputBox("Enter design id to open
...")
        LoadDesign(designid)
    End Sub
```

7. One final change before we build our solution: go to **Form Designer**, double-click on the `lblDesignID` label, and call the `LoadDesign` method. This will reload all excavations every time we click on the **Design ID** label and will prove useful in the coming pages. This will be our refresh button:

```
    Private Sub lblDesignID_Click(sender As Object, e As
EventArgs) Handles lblDesignID.Click
        LoadDesign(txtDesignID.Text)
    End Sub
```

8. Build your solution and run `yharnam.mxd`.

9. Open **Design Manager** and click on **Open Design**.

10. Type in a design ID that you have already created before and check your result. In my case, design ID 3 shows the following excavations:

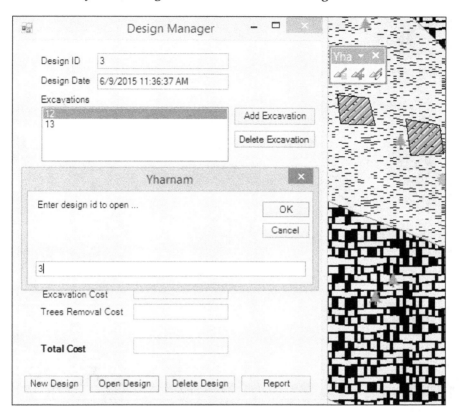

11. Click on **Add Excavation** and add one more excavation to this design.

12. Click on the **Design ID** label to refresh the new excavation, as illustrated in the following screenshot:

13. Close ArcMap and choose not to save the changes.

Loading trees

To load the underlying trees of all excavations, we are required to run the spatial query on each excavation and populate those trees in that excavation:

1. If necessary, open Visual Studio Express in administrator mode; we need to do this since our project is actually writing to the registry this time, so it needs administrator permissions. To do that, right-click on **Visual Studio** and click on **Run as administrator**.

2. Go to **File**, then click on **Open Project,** browse to the Yharnam project from the C:\ArcGISByExample\yharnam\Code, and click on **Open**.

3. Open frmDesignManager.vb in the **Code** view.

4. In the LoadDesign method, add the following lines to the Excavation loop:

```
Public Sub LoadDesign(designid As Long)



        Dim pWorkspace As IWorkspace = getYharnamWorkspace()
        Dim pTable As ITable = getDesignTable(pWorkspace)
        Dim pQFilter As IQueryFilter = New QueryFilter
        pQFilter.WhereClause = "DESIGNID = " & designid
        Dim pCursor As ICursor = pTable.Search(pQFilter,
False)
        Dim pRow As IRow = pCursor.NextRow

        If pRow Is Nothing Then
            MsgBox("Design was not found!")
            Exit Sub
        End If

        txtDesignID.Text = pRow.Value(pRow.Fields.
FindField("DESIGNID"))
        txtDesignDate.Text = pRow.Value(pRow.Fields.
FindField("DESIGNDATE"))

        Dim pFeatureClass As IFeatureClass = getExcavationFeat
ureClass(pWorkspace)
        Dim pFCursor As IFeatureCursor = pFeatureClass.
Search(pQFilter, False)
        Dim pFeature As IFeature = pFCursor.NextFeature
        lstExcavations.Items.Clear()
        lstTrees.Items.Clear()
        Do Until pFeature Is Nothing
            lstExcavations.Items.Add(pFeature.OID)


        Dim pTreeFeatureClass As IFeatureClass = getTreesF
eatureClass(pWorkspace)
        Dim pSFilter As ISpatialFilter = New SpatialFilter
        pSFilter.Geometry = pFeature.Shape
        pSFilter.SpatialRel = esriSpatialRelEnum.
esriSpatialRelIntersects
```

```
                        Dim pFeatureCursor As IFeatureCursor =
pTreeFeatureClass.Search(pSFilter, False)
                        Dim pTreeFeature As IFeature = pFeatureCursor.
NextFeature

                        Do Until pTreeFeature Is Nothing
                            lstTrees.Items.Add(pTreeFeature.OID)
                            pTreeFeature = pFeatureCursor.NextFeature
                        Loop

                        pFeature = pFCursor.NextFeature
                    Loop


        End Sub
```

5. Now we will create two variables, `TotalTreeCount`, which will sum the total number of trees in all excavations, and `TotalArea`, which is the sum of all excavation areas. We will then display these two variables on the form. Write the following code to the `LoadDesign` method to do so:

```
        Public Sub LoadDesign(designid As Long)

        . . . .
        . . . .
        . . . .


                        Dim pFeatureClass As IFeatureClass = getExcavationFeat
ureClass(pWorkspace)
                        Dim pFCursor As IFeatureCursor = pFeatureClass.
Search(pQFilter, False)
                        Dim pFeature As IFeature = pFCursor.NextFeature
                        lstExcavations.Items.Clear()
                        lstTrees.Items.Clear()
                        Dim totalTreeCount As Long = 0
                        Dim dtotalArea As Double = 0
```

```
        Do Until pFeature Is Nothing
            lstExcavations.Items.Add(pFeature.OID)

            Dim pArea As IArea = pFeature.Shape
            dtotalArea = dtotalArea + Math.Abs(pArea.Area)

            Dim pTreeFeatureClass As IFeatureClass = getTreesF
eatureClass(pWorkspace)
            Dim pSFilter As ISpatialFilter = New SpatialFilter
            pSFilter.Geometry = pFeature.Shape
            pSFilter.SpatialRel = esriSpatialRelEnum.
esriSpatialRelIntersects

            Dim pFeatureCursor As IFeatureCursor =
pTreeFeatureClass.Search(pSFilter, False)
            Dim pTreeFeature As IFeature = pFeatureCursor.
NextFeature

            Do Until pTreeFeature Is Nothing
                lstTrees.Items.Add(pTreeFeature.OID)
totalTreeCount = totalTreeCount + 1
                pTreeFeature = pFeatureCursor.NextFeature
            Loop

            pFeature = pFCursor.NextFeature
        Loop

        txtTotalArea.Text = Math.Round(dtotalArea, 2) & "m2"
        txtTreeCount.Text = totalTreeCount

    End Sub
```

6. Build your solution and run yharnam.mxd.

7. Open **Design Manager** and click on **New Design**.

8. Click on **Add Excavation** and draw three excavations around multiple trees.

9. Click on the **Design ID** label to refresh the design and see your excavations, total number of trees, and total area, as illustrated in the following screenshot:

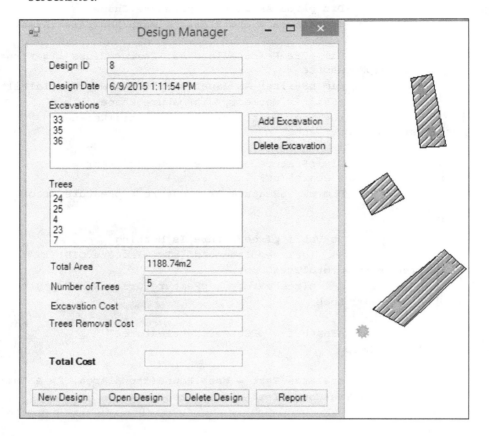

10. Close ArcMap and choose not to save the changes.

Calculating the design cost

Now that we have all the excavations, we can calculate the cost of each one and sum them to get the total cost. Thanks to our excavation calculator that we have developed in the previous chapter, this job is now easy:

1. If necessary, open Visual Studio Express in administrator mode; we need to do this since our project is actually writing to the registry this time, so it needs administrator permissions. To do that, right-click on **Visual Studio** and click on **Run as administrator**.

2. Go to **File**, then click on **Open Project**, browse to the `Yharnam` project from `C:\ArcGISByExample\yharnam\Code`, and click on **Open**.

3. Modify your `LoadDesign` method located in `frmDesignManager.vb` to call `ExcavationCalculator` for every excavation, and sum these values and display each in its corresponding text box:

```
Public Sub LoadDesign(designid As Long)

    .   .   .   .
    .   .   .   .
    .   .   .   .

Dim totalTreeCount As Long = 0
            Dim dtotalArea As Double = 0
            Dim dtreeRemovalcost As Double = 0
            Dim dExcavationCost As Double = 0
            Dim dTotalCost As Double = 0

            Do Until pFeature Is Nothing
                lstExcavations.Items.Add(pFeature.OID)

                Dim pExcavationCalculator As New ExcavationCostCal
culator(pFeature)
                dtreeRemovalcost = dtreeRemovalcost +
pExcavationCalculator.TreesRemovalCost
                dExcavationCost = dExcavationCost +
pExcavationCalculator.SoilRemovalCost()

                Dim pArea As IArea = pFeature.Shape
                dtotalArea = dtotalArea + Math.Abs(pArea.Area)

                Dim pTreeFeatureClass As IFeatureClass = getTreesF
eatureClass(pWorkspace)
                Dim pSFilter As ISpatialFilter = New SpatialFilter
                pSFilter.Geometry = pFeature.Shape
                pSFilter.SpatialRel = esriSpatialRelEnum.
esriSpatialRelIntersects

                Dim pFeatureCursor As IFeatureCursor =
pTreeFeatureClass.Search(pSFilter, False)
                Dim pTreeFeature As IFeature = pFeatureCursor.
NextFeature

                Do Until pTreeFeature Is Nothing
                    lstTrees.Items.Add(pTreeFeature.OID)
                    totalTreeCount = totalTreeCount + 1
                    pTreeFeature = pFeatureCursor.NextFeature
                Loop
```

```
                    pFeature = pFCursor.NextFeature
            Loop

            txtTotalArea.Text = Math.Round(dtotalArea, 2) & "m2"
            txtTreeCount.Text = totalTreeCount
            txtTreeRemovalCost.Text = "$" & Math.
Round(dtreeRemovalcost, 2)
            txtExcavationCost.Text = "$" & Math.
Round(dExcavationCost, 2)
            txtTotalCost.Text = "$" & Math.Round(dtreeRemovalcost
+ dExcavationCost, 2)
      End Sub
```

4. Build your solution. If it fails, make sure you have run the solution as administrator.

5. Run `yharnam.mxd`.

6. Run **Design Manager**, click on **Open Design**, specify a design ID that you previously have saved, and take a look at the calculating costs, as shown in the following screenshot:

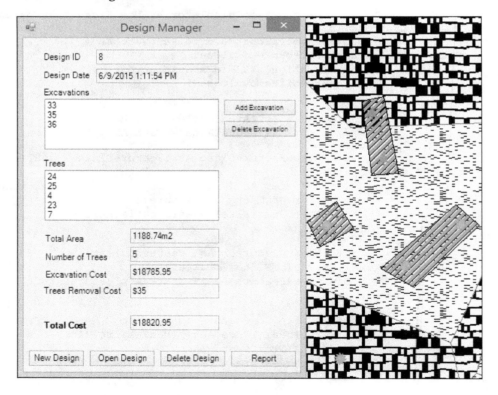

7. Close ArcMap and choose not to save the changes.

Using the filter map to show only design excavation

We have almost completed our open design algorithm. Now we only have to add a final touch to filter the map to show the excavations that belong to the design we are opening. Follow these steps:

1. Make sure your `Yharnam` project is opened as administrator.

2. We will need to get hold of the `Excavation` layer, which is the second layer on the map, and then we will add a filter to show only excavations with the design ID we are opening. Go to your `LoadDesign` method in `frmDesignManager` and add the following code just before the end:

    ```
    . . . .
    . . . .
    . . . .
            Dim pDoc As IMxDocument = _application.Document
            Dim pLayerDef As IFeatureLayerDefinition = pDoc.
    FocusMap.Layer(1)
            pLayerDef.DefinitionExpression = "DESIGNID=" &
    designid
            pDoc.ActiveView.Refresh()
    End Sub
    ```

3. When we close the form, we want to make sure that we've closed the design and restored the layer to its original state. Write the following code in `FormClosed`:

    ```
        Private Sub frmDesignManager_FormClosed(sender As Object, e As
    Windows.Forms.FormClosedEventArgs) Handles Me.FormClosed
            _application.CurrentTool = Nothing
            Dim pDoc As IMxDocument = _application.Document
            Dim pLayerDef As IFeatureLayerDefinition = pDoc.FocusMap.
    Layer(1)
            pLayerDef.DefinitionExpression = ""
            pDoc.ActiveView.Refresh()
        End Sub
    ```

4. Build your solution. If it fails, make sure you have run the solution as administrator.

5. Run `yharnam.mxd`.

6. Run **Design Manager**, click on **Open Design**, specify a design ID that you have previously saved, and take a look at the map. You will see that you can now only see the excavations for that design.

7. Close ArcMap and choose not to save the changes.

Deleting designs

In this section, we will continue working on our design manager; only few functionalities are missing and among them are delete excavation and delete design.

Deleting an excavation

To delete an excavation from an existing design, we will need to select it from the list and click on the **Delete Excavation** button. Then we are required to call the LoadDesign method to reload everything:

1. If necessary, open Visual Studio Express in administrator mode; we need to do this since our project is actually writing to the registry this time, so it needs administrator permissions. To do that, right-click on **Visual Studio** and click on **Run as administrator**.

2. Go to **File**, then click on **Open Project**, browse to the Yharnam project from the C:\ArcGISByExample\yharnam\Code, and click on **Open**.

3. Double-click on frmDesignManager.vb to view the form designer.

4. Double-click on the **Delete Excavation** button to generate the btnDeleteExcavation_Click method.

5. The lstExcavation.SelectedItem variable will give us the selected excavation object ID. We will use it to query that feature, delete it, and refresh the design. This should all be done through an editing session, as follows:

```
Private Sub btnDeleteExcavation_Click(sender As Object, e As
EventArgs) Handles btnDeleteExcavation.Click
        Dim lExcavationOID As Long = lstExcavations.SelectedItem
        Dim pWorkspaceEdit As IWorkspaceEdit =
getYharnamWorkspace()
        pWorkspaceEdit.StartEditing(True)
        pWorkspaceEdit.StartEditOperation()

        Dim pFeatureClass As IFeatureClass = getExcavationFeatureC
lass(pWorkspaceEdit)

        Dim pFeature As IFeature = pFeatureClass.
GetFeature(lExcavationOID)
        pFeature.Delete()
        pFeature.Store()
```

```
pWorkspaceEdit.StopEditOperation()
pWorkspaceEdit.StopEditing(True)

LoadDesign(txtDesignID.Text)
End Sub
```

6. Build your solution and run `yharnam.mxd`.

7. Open a design with multiple excavations and select one of the excavations from `lstExcavations`, then click on **Delete Excavation**. You will see that the excavation has been removed from both the list and the map.

8. Close ArcMap and choose not to save the changes.

Deleting a design

Deleting a design is a bit more challenging than deleting a single excavation. A design requires us to delete the design record and all underlying excavations that belong to that design. Follow these steps to learn how to delete a design:

1. If necessary, open the `Yharnam` project as administrator.

2. Double-click on `frmDesignManager.vb` to view the form designer.

3. Double-click on the **Delete Design** button to generate the `btnDeleteDesign_Click` method.

4. First we are required to start an edit session, then get the table and delete the design record as follows:

```
Private Sub btnDeleteDesign_Click(sender As Object, e As
EventArgs) Handles btnDeleteDesign.Click
    Dim pWorkspaceEdit As IWorkspaceEdit =
getYharnamWorkspace()
    pWorkspaceEdit.StartEditing(True)
    pWorkspaceEdit.StartEditOperation()

    Dim pTable As ITable = getDesignTable(pWorkspaceEdit)
    Dim pRow As IRow = pTable.GetRow(txtDesignID.Text)
    pRow.Delete()
    pRow.Store()

    pWorkspaceEdit.StopEditOperation()
    pWorkspaceEdit.StopEditing(True)

End Sub
```

5. To delete all excavation features, we can use the same code we used for **Delete Excavation** as follows:

```
Private Sub btnDeleteDesign_Click(sender As Object, e As
EventArgs) Handles btnDeleteDesign.Click
        Dim pWorkspaceEdit As IWorkspaceEdit =
getYharnamWorkspace()
        pWorkspaceEdit.StartEditing(True)
        pWorkspaceEdit.StartEditOperation()

        Dim pTable As ITable = getDesignTable(pWorkspaceEdit)
        Dim pRow As IRow = pTable.GetRow(txtDesignID.Text)
        pRow.Delete()
        pRow.Store()


        Dim pFeatureClass As IFeatureClass = getExcavationFeatureC
lass(pWorkspaceEdit)
        Dim pQFilter As IQueryFilter = New QueryFilter
        pQFilter.WhereClause = "DESIGNID = " & txtDesignID.Text
        Dim pFCursor As IFeatureCursor = pFeatureClass.
Update(pQFilter, False)
        Dim pFeature As IFeature = pFCursor.NextFeature

        Do Until pFeature Is Nothing
            pFeature.Delete()
            pFeature.Store()

            pFeature = pFCursor.NextFeature
        Loop

        pWorkspaceEdit.StopEditOperation()
        pWorkspaceEdit.StopEditing(True)
        Me.Close()
    End Sub
```

6. Build your solution run yharnam.mxd.

7. Open a design that has some existing excavations and click on **Delete Design** to remove it from the geodatabase. This will close the form. Refresh the map to verify that the excavations have been deleted.

8. Close ArcMap and choose not to save the changes.

Generating the excavation design report

Reporting is a very important functionality in any application. The ability for the software to generate reports that can later be e-mailed, printed, and published on the Web, can prove the visibility of the system. In this section, we will learn how to generate an HTML report for our excavation design:

1. If necessary, open the `Yharnam` project as administrator.

2. Double-click on `frmDesignManager.vb` to view the form designer.

3. Double-click on the **Report** button to generate the `btmReport_Click` method.

4. This is a straightforward functionality. We will simply build an HTML file and write all form details into the file as follows:

```
Private Sub btnReport_Click(sender As Object, e As EventArgs)
Handles btnReport.Click
        Dim sHTML As String = ""

        sHTML &= "<html><body><br>"
        sHTML &= "<h1>Yharnam Excavation Planning Manager:</h1>"
        sHTML &= "<b>DesignID:</b> " & txtDesignID.Text & "<br>"
        sHTML &= "<b>Design Date:</b> " & txtDesignDate.Text &
"<br>"
        sHTML &= "<b>Total Area:</b> " & txtTotalArea.Text &
"<br>"
        sHTML &= "<b>Number of Trees:</b> " & txtTreeCount.Text &
"<br>"
        sHTML &= "<b>Excavation Cost:</b> " & txtExcavationCost.
Text & "<br>"
        sHTML &= "<b>Tree Removal Cost:</b> " &
txtTreeRemovalCost.Text & "<br>"
        sHTML &= "<br><br>"
        sHTML &= "<h3>Total Cost:</h3>" & txtTotalCost.Text

        sHTML &= "</body></html>"
        Dim sPath As String = "C:\ArcGISByExample\yharnam\Report.
html"
        IO.File.WriteAllText(sPath, sHTML)
        Process.Start(sPath)

    End Sub
```

5. Build your solution and run `yharnam.mxd`.

6. Open the Design Manager and click on the **Open Design**, type a design ID of one of your designs.

7. Click on the **Report** button to generate the report, as shown in the following screenshot:

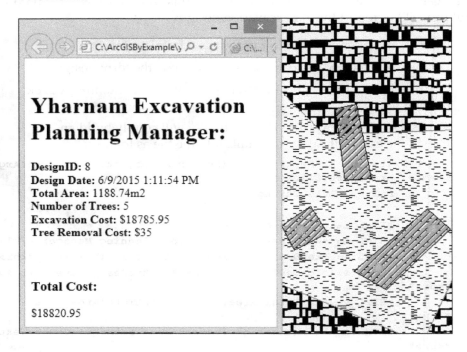

8. Close ArcMap and choose not to save the changes.

Searching for the design

We have reached the final stage of our Excavation Planning Manager, which is to list and search for designs. For that, we will be adding a new button and a new form to list all the designs. You don't have to implement this task as I have already added it in the final code B04847_10_Files\yharnam\FinalCode\:

1. Use the final code to compile and run Yharnam project.

2. You will see a new button called **Search Design**, click on it to open the **Search Design** form.

3. Select a **From** and **To** date to filter designs created between those two dates, as illustrated in the following screenshot:

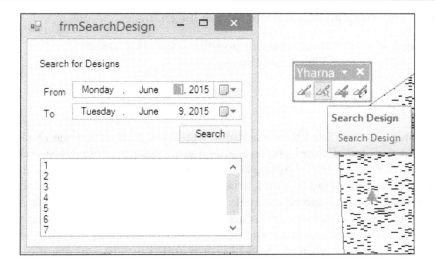

4. Click on a design ID from the list to open the Design Manager for that particular design. This is shown in the next screenshot:

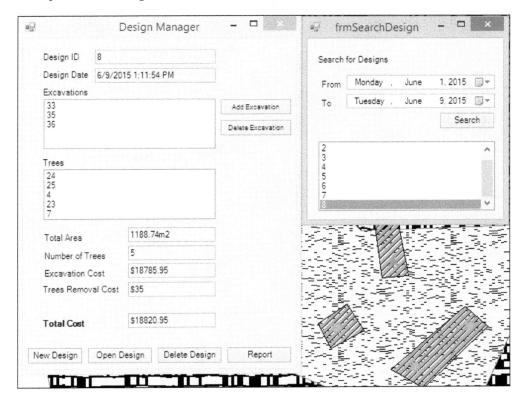

5. Close ArcMap and choose not to save the changes.

This is the end of this chapter; you can find the latest code under `B04847_10_Files\yharnam\FinalCode\`.

Summary

In this chapter, you finalized the Excavation Manager. You added the Design Manager form, which allowed you to manage your designs, create, search, edit, and delete. Throughout this chapter, you added new functionality to the design manager until you finalized it. You learned how to call an existing tool and use it to draw on the map to create an excavation. You learned how to generate a design report and, finally, you managed to search and open existing designs that you had already saved. This way you have enabled Yharnam construction engineers to create multiple designs on top of the ArcGIS platform and compare the cost of executing each one to select optimum option.

This is the end of this book, but it is only the beginning of great, potential applications that you will develop using the skill set you acquired during the course of this journey.

Index

coordinate system
  about 43-47
  URL 44

D

design manager, excavation design
  button, creating 201, 202
  creating 201
  form, creating 202-204
  multiple excavations, saving 207-210
  new design, creating 204-206
Dynamic Link Library (DLL) 12

E

excavation cost calculator
  ExcavationCostCalculator class,
    creating 182, 183
  preparing 181, 182
excavation design
  deleting 222-224
  design cost, calculating 218-220
  design manager, creating 201
  design table, preparing 200, 201
  excavation, deleting 222, 223
  excavations, loading 211-214
  existing designs, opening 210
  LoadDesign method, creating 210
  map, filtering 221
  reporting 225, 226
  searching 226-228
  trees, loading 214-218
excavation planning manager
  about 18, 19
  excavation cost calculator,
    preparing 181, 182
  excavation editor tool, adding 164-166
  excavation features, creating 168
  excavation information, editing 175-179
  excavation information, viewing 175-179
  excavation manager toolbar,
    adding 166, 167
  excavation tool, adding 163
  final cost, estimating 194-196
  geometries, converting into
    features 172-174

geometries, drawing on map 168-171
  number of trees, calculating 189, 190
  project, preparing 160
  soil type removal cost, calculating 184-188
  tree removal cost, calculating 189-193
  Yharnam geodatabase, preparing 160, 161
  Yharnam map, preparing 160, 161
  Yharnam project, preparing 161, 162
external GPS point coordinates
  mapping 75, 76
  reading 73, 74

G

geodatabase
  editing 159, 160
  workspace, creating 103
geographic information systems (GIS) 1
Global Positioning System (GPS) 16
graphic elements 37

I

**Integrated Development Environment
  (IDE) 13**

L

licenses, ArcGIS for Desktop
  Advanced 3
  Basic 3
  Standard 3
  URL 3

M

map layers 9-11

O

ObjectID 34

R

real-time cell phone simulator button add-in
  adding 66-69
  map point, creating 70-72

regions, restaurants mapping application
   geodatabase, connecting to 136-138
   populating 138-143
   querying 135
   restaurants, finding 143
   restaurants, populating 144-149
   spatial queries 144

relationships
   about 113
   ratings table 114-116
   reviews table 114-116

restaurants
   filtering 153-156
   filtering, on map 131-133
   finding, in region 143
   highlighting 124
   populating, in region 144-149
   restaurants subtypes, querying 102
   searching, in subtype 108-111

restaurants mapping application
   about 17, 18
   ArcGIS Display object, using 124-128
   average rating, calculating 122, 123
   features, highlighting 128-131
   geodatabase workspace 103-105
   ratings table 114-117
   real-time search 153-156
   records, retrieving 117-121
   restaurants mapping toolbar, adding 94-97
   restaurants viewer button, adding 97-100
   restaurants viewer button, assigning to
      toolbar 101, 102
   reviews table 114-117
   search textbox, adding in toolbar 150-152
   subtypes, populating 106, 107

S

search textbox
   adding, in toolbar 150-153
signal maneuvering
   enabling 77
   enabling, with timer 80-83
   GPS file, loading 77-80
Software as a Service (SaaS) 2

spatial queries 144
spatial reference
   about 5
   URL 5
Standard license, ArcGIS for Desktop
   about 3
   URL 3

T

toolbar
   search textbox, adding 150-153
topological operators 37
towers range, cell tower analysis tool
   displaying 34
   drawing 39-41
   drawing, based on attribute value 38
   features, querying 34-37
   graphic elements 37, 38
   range attribute, drawing 39
   topological operators 37, 38

V

Visual Basic for Applications (VBA) 12, 21

W

workspace
   about 103
   creating 103-105
   URL 103
workspace factory object 103

Thank you for buying
ArcGIS By Example

About Packt Publishing

Packt, pronounced 'packed', published its first book, *Mastering phpMyAdmin for Effective MySQL Management*, in April 2004, and subsequently continued to specialize in publishing highly focused books on specific technologies and solutions.

Our books and publications share the experiences of your fellow IT professionals in adapting and customizing today's systems, applications, and frameworks. Our solution-based books give you the knowledge and power to customize the software and technologies you're using to get the job done. Packt books are more specific and less general than the IT books you have seen in the past. Our unique business model allows us to bring you more focused information, giving you more of what you need to know, and less of what you don't.

Packt is a modern yet unique publishing company that focuses on producing quality, cutting-edge books for communities of developers, administrators, and newbies alike. For more information, please visit our website at www.packtpub.com.

Writing for Packt

We welcome all inquiries from people who are interested in authoring. Book proposals should be sent to author@packtpub.com. If your book idea is still at an early stage and you would like to discuss it first before writing a formal book proposal, then please contact us; one of our commissioning editors will get in touch with you.

We're not just looking for published authors; if you have strong technical skills but no writing experience, our experienced editors can help you develop a writing career, or simply get some additional reward for your expertise.

ArcGIS for Desktop Cookbook

ISBN: 978-1-78355-950-3   Paperback: 372 pages

Over 60 hands-on recipes to help you become a more productive ArcGIS for Desktop user

1. Learn how to use ArcGIS Desktop to create, edit, manage, display, analyze, and share geographic data.

2. Use common geo-processing tools to select and extract features.

3. A guide with example-based recipes to help you get a better and clearer understanding of ArcGIS Desktop.

Learning ArcGIS Geodatabases

ISBN: 978-1-78398-864-8   Paperback: 158 pages

An all-in-one start up kit to author, manage, and administer ArcGIS geodatabases

1. Covers the basics of building Geodatabases, using ArcGIS, from scratch.

2. Model the Geodatabase to an optimal state using the various optimization techniques.

3. Packed with real-world examples showcasing ArcGIS Geodatabase to build mapping applications in web, desktop, and mobile.

Please check **www.PacktPub.com** for information on our titles

Administering ArcGIS for Server

ISBN: 978-1-78217-736-4   Paperback: 246 pages

Installing and configuring ArcGIS for Server to publish, optimize, and secure GIS services

1. Configure ArcGIS for Server to achieve maximum performance and response time.

2. Understand the product mechanics to build up good troubleshooting skills.

3. Filled with practical exercises, examples, and code snippets to help facilitate your learning.

Building Web Applications with ArcGIS

ISBN: 978-1-78355-295-5   Paperback: 138 pages

Build an engaging GIS Web application from scratch using ArcGIS

1. Learn how to design, build, and run high performance and interactive applications with the help of ArcGIS.

2. Incorporate ArcGIS for Server services to allow end users to visualize, query, and edit GIS data using the ArcGIS JavaScript APIs.

3. Step-by-step tutorial that teaches you how to design and customize a GIS web application from scratch.

Please check **www.PacktPub.com** for information on our titles

www.ingramcontent.com/pod-product-compliance
Lightning Source LLC
Chambersburg PA
CBHW060538060326
40690CB00017B/3527